Adam Spencer's Book of Numbers

ADAM SPENCER'S
Book of
Numbers

Four Walls Eight Windows

New York

© 2004 Adam Spencer

Published in the United States by:
Four Walls Eight Windows
39 West 14th Street, room 503
New York, N.Y., 10011

Visit our website at http://www.4w8w.com
First U.S. printing January 2004.

Published in arrangement with Penguin Books Australia.

Library of Congress Cataloging-in-Publication Data on file.

ISBN 1-56858-289-7
10 9 8 7 6 5 4 3 2 1

Typesetting by Pracharak Technologies
Printed in Canada on 100% recycled paper

CONTENTS

INTRODUCTION

Welcome to this bizarre walk through the mathematics and pop culture of the numbers from 1 to 100. At this stage you're entitled to ask 'Where on earth did the idea for this come from?' Here is my best attempt to answer your very valid question.

First up, I love numbers. And I say that with pride. Being blown away by the world of maths isn't just a trip for 2000-year-old professors with out-of-control hair, nor for cardigan-clad and beige-bottomed bozos. Numbers are everywhere around us and the little tackers are fascinating. Hidden patterns, relationships and bizarre facts run through the world of numbers and many of them are amazing. This book shows you some of those neato facts in a way you can understand and explains some of the mathematical problems that have fascinated and confused even the best mathematicians for centuries.

Best of all, the book is entirely self-contained. Just from reading the stuff here you'll be able to explain hundreds of interesting facts about the world of numbers to other people, armed with a mere pen and paper. Imagine the number of guests you can stun at dinner parties or social gatherings as you prove that if more than 23 people are in a room, odds are there'll be a common birthday.

Hopefully, by reading the facts I've assembled here, you'll start to share my love affair with numbers. Okay, that's a bit excessive, Adam, but at least you might start to realise how fascinating the world of numbers can be. Maths isn't something to be feared or looked back on with terror from your high-school days. It's an amazing and entertaining world – and with luck, this book will be your first taste of the fabulous all-you-can-eat buffet that awaits.

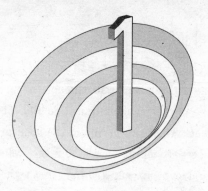

It's not strictly correct to say that 1 is the 'first number', because there are negative numbers and 0 and other numbers that are 'less than' 1. But it is certainly the first number most of us encounter and probably the first number of which anyone ever thought. Now, who that guy or girl was is unknown, but their realising that we can have 1 of something or 2 of them or small bits of things that aren't really 1 thing is where this whole wacky caper called mathematics started.

Some things only happen once. Snails have sex only once in their lives. But see ➤12 to find out why, for a snail, the news isn't all bad.

We all know that $1 \times 1 = 1$ and $1 \times 2 = 2$. In fact, $1 \times n = n$ when n is any number. As a result 1 is called the **multiplicative identity** (when you multiply anything by 1, its identity remains unchanged).

Once is how often every citizen of Kentucky is required by law to have a bath each year. But see ➤71 for good reasons why Kentucky law-makers should get tough on soap dodgers.

Kenny only survived 1 episode of *South Park* in the first 2 seasons. You bastards! Take a look at ➢39.

QUIZ QUESTION

What is the name of the 1st official episode of *South Park*?

There's a 1st for everything. The 1st toilet ever shown on television was on *Leave it to Beaver*. The 1st couple to be seen sharing a bed on a TV show were Lily and Herman Munster.

Ahem: 1 + 1 = 2. Before you rush into 'Yeah, thanks for that, Adam,' this is in fact an incredibly profound statement that, if you're being strict about it, takes tremendous effort to prove.

Even numbers are defined as numbers divisible by 2. Two is the 1st even number.

Now, $15 = 3 \times 5$ and $77 = 7 \times 11$, but 2 can only be written as $2 = 1 \times 2$. Because of this, 2 is called a **prime number**. A prime number is a number that can't be written as any 2 whole numbers multiplied together except 1 and itself. Two is the only even prime number, because every other even number is divisible by 2, and so can't be prime.

The early Greeks weren't sure whether 2 was a number at all, because it has a beginning and an end, but no middle. Being an even number it was considered female, the 1st **female number**, in fact.

Yin and Yang are the 2 fundamental principles in the Chinese concept depicting the duality of life. Also, the 2 small sticks used in the I Ching are a binary system. Lots of religions are dualistic, like Zoroastrianism, the ancient Iranian religion with its Ahura

Mazda, the god of light and goodness, and Ahriman, the dark, evil principle.

Of course, 2 is the basis of the **binary system**. In 1725 Basile Bouchon invented a device consisting of a roll of perforated paper punched with holes that could control the threads on a mechanical loom. The same idea was used in the pianola. Then Charles Babbage used the system in his Analytical Engine (also see ➤79). Thanks to binary notation, today's analytical engines can do all kinds of things, even make green stuff: ask Bill Gates.

'You Only Live Twice' was sung by Nancy Sinatra and was the theme music to the James Bond movie of the same name. 'Two Tribes', released in 1984, was Frankie Goes to Hollywood's 2nd single. Like their 1st, 'Relax', it went better than Number 2 on the charts, getting to Number 1.

It takes 2 to tango, and 8 to quadrille, but you only need 1 buffoon in a cowboy hat to start people linedancing.

Three is a **Fibonacci number**. The Fibonacci numbers are: 1, 1, 2, 3, 5, 8, 13, 21 . . . Can you see the pattern? You add 2 numbers in a row to get the next number. The Fibonacci numbers are among the most famous and often-used numbers in mathematics. For more information, have a look at ➤8.

Three places claim to be home to the biggest egg in the world: Mentone, Indiana; Newberry, South Carolina; Winlock, Washington. No surprise, really, that they're all in the USA . . .

Our spatial world has 3 dimensions (length, height and width) and we divide time into 3 (past, present and future). There are 3 primary colours (red, blue and yellow) and 3 states of matter (solid, liquid and gas). Hinduism has Brahma the Creator, Shiva the Destroyer and Vishnu the Sustainer; and Buddhism has the 3 classifications of teachings.

There is an easy way to check if a number is divisible by 3. If its digits add up to a number divisible by 3, then it can be divided by 3. There are no exceptions. So, we know that 39 123 is divisible by 3 because 3 + 9 + 1 + 2 + 3 = 18 and 18 = 3 × 6. In case you're wondering, 39 123 = 13 041 × 3.

QUIZ QUESTION

Which of these 5 numbers is divisible by 3?
281, 354, 962, 3742, 138 624 147

The basis of harmony in western music is the chord, the simplest of which is the triad. A basic triad consists of a root note, a 3rd and a 5th: 3 notes.

QUIZ QUESTION

Three different Mr Hats have appeared in *South Park*.
What are they?

Three is a popular number for girl pop groups. Think of Martha and the Vandellas, the Ronettes, the Pointer Sisters, the Supremes, TLC, Salt-N-Pepa – not forgetting that great 1980s girl group, Bananarama, and their immortal hit 'Robert De Niro's Waiting'. Christianity also has as its Big Three the Father, Son and Holy Spirit. Any link between this and Bananarama is tenuous at best.

Because $4 = 2 \times 2$, it is the first **square number**. We say that 4 is '2 squared' and write this $4 = 2^2$. To explain this notation, if you were to write $5 \times 5 \times 5$ you could also write it as 5^3.

Four is the perfect number for the Sioux of North America: they have 4 groups of gods, 4 species of animal, 4 ages of the human being and their medicine men advised people to do things in groups of 4.

A strand of DNA is just a combination of 4 proteins – adenine, cytosine, guanine and thymine.

Four is the 1st **composite number**, that is, a number that can be divided by numbers other than itself and 1. Pythagoreans called numbers divisible by 4 'even-even', so they associated 4 with harmony and justice.

Some cultures' counting systems are **base 4** (see ➤10 if you're not sure what that means): 4 fingers equal a hand and 4 hands equal a foot.

The 4 phases of the moon were the 1st means of keeping time, and the 4 cardinal points gave order to space. Albert

Einstein's space–time is 4-dimensional, although later theorists have suggested that 4 dimensions are not enough.

In Mayan thought, everything is related to the 4 cardinal points, which were represented by a cross that touched the 4 horizons. In Christianity the 4 directions are represented by the 4 angels, which stand for God's power extending to the entire world. The cross obtained by connecting the 4 cardinal points was, in Ancient Egypt, the hieroglyph for 'immortality'.

Among Indian gods, Brahma has 4 heads to symbolise the 4 directions of the world, while Shiva has 4 arms with which he destroys and recreates the world in his dance.

There are 4 prophets in the Old Testament and 4 rivers of paradise, which Christians believe prefigure the 4 Gospels of the New Testament. There are 4 sacred books in Islam: the Torah, Psalms, Gospel and Quran.

Chess players might like to know that the number of different ways of playing just the 1st 4 moves on each side in a game is 318 979 564 000.

The national anthems of Japan, Jordan and San Marino each have only 4 lines, which makes them a lot easier to remember than certain other national anthems.

Anyone who's ever blown a few bucks playing Tetris has seen the 5 possible **tetrominoes**. A tetromino is a shape made by joining together 4 squares edge to edge, like this:

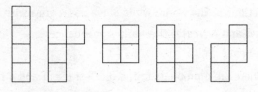

QUIZ QUESTION

How many basic shapes can you make from 5 squares? These are called *pentominoes.* **The squares must be joined edge to edge, and you can't count 2 shapes if they match when rotated and reflected.**

In western and eastern cultures 5 is often the number of love, being an indivisible combination of the masculine 3 and the feminine 2. The pentagram is called the lovers' knot in English tradition, and 5 is the number of Venus, the goddess of love. Sexy!

Legend has it that Hippasos, the guy who discovered the **pentagondodecahedron**, which is the 5th geometrical body and consists of 12 regular pentagons, was drowned by the Pythagoreans for it. A pretty tough call for discovering a shape.

The Pentagon in Arlington, Virginia, containing most of the US Defense Department offices, has 5 sides, 5 storeys and 5 acres (2.025 hectares) in the middle. Now that's really sexy!

There are 5 notes in the pentatonic scale, which Celtic music is often based on.

Five members is the preferred number for all-boy harmony groups. Consider Backstreet Boys, Boyzone, 'N Sync, 5ive, Take That, New Kids on the Block etc. etc. The allocation of talent across the 5 members is traditionally 2 fantastic breakdancers; 1 guy who looks great in slacks; 1 moody, brooding loner (complete with nose ring and goatee beard); and 1 who can actually sing.

Six is a **perfect number**. Perfect numbers are numbers that are equal to the sum of their factors (not including themselves). So 6 can be written as 1×6 and 2×3. When we add the factors, $1 + 2 + 3 = 6$.

QUIZ QUESTION

What are the next 2 perfect numbers after 6?
Hint: the 2nd perfect number is between 20 and 40, and if you're feeling gutsy, the 3rd one is between 490 and 500. Good luck.

The 6-sided figure, the **hexagon**, is an important building principle in nature: beehives and snowflakes are constructed of hexagons.

There are 6 notes in the whole-tone scale, which is found in some eastern music. It's comprised of whole tones, in case you were wondering.

In the hit TV show of 1976–81 *Charlie's Angels*, there were 6 angels. Well, there were only 3 at a time, but there were 6 different angels over the 5 years that the show ran. They were: Kate Jackson, aka Sabrina Duncan; Farrah Fawcett, aka Jill

Munroe; Jaclyn Smith, aka Kelly Garrett; Cheryl Ladd, aka Kris Munroe; Shelley Hack, aka Tiffany Welles; and Tanya Roberts, aka Julie Rogers. Why they never got all 6 together to take on some serious mega-villains I don't know.

Seven is a prime number, but because it is of the form $2^p - 1$ ($7 = 2^3 - 1$), where p is a prime number, it is called a **Mersenne prime**, after Father Marin Mersenne, a mathematician, philosopher and friend of Descartes. Mersenne primes are very rare indeed. Until recently only 37 had been discovered. Check out ➤31 for the 38th.

QUIZ QUESTION

Find the next 2 Mersenne primes after 7. Remember, they must be of the form $2^p - 1$, where p is prime.

Lust, pride, anger, envy, sloth, avarice and gluttony. Sounds like fun, eh? That's because the deadly sins are a **heptad**, of course: a series of 7 things. Other heptads include the colours of the rainbow (ROYGBIV – see also ➤40), the number of layers in the Mayan sky, and the number of planets in the Babylonian heavens. Seven is also the number of Mary's joys and of her sorrows in Christianity, and the number of hills in Rome. And that's just for starters (the 7-year itch, 7 years' bad luck, the 7 lives of an Iranian cat . . .).

Seven is a popular word in movie titles: *Seven* (1979), *Se7en* (1995), *The Magnificent Seven*, *Seven Brides for Seven Brothers*, *The*

Seven Brothers Meet Dracula, *Seven Doors to Death* and *The Seven Samurai* to name just a few. But one you won't forget is *The Seven Magnificent Gladiators* (1984), starring the Incredible Hulk and some other, lesser-known Thespians.

Vulcans, such as Spock in *Star Trek*, are sexually active just once every 7 years. Logical? Well, not to me.

Eight is a Fibonacci number. Fibonacci, or Leonardo of Pisa, wrote a famous book called *Liber abaci* in 1202, which included the 'rabbit problem'. If 2 rabbits give birth to a new pair each month, but the new pair don't breed until they're a month old, and the rabbits never die, how many pairs are alive each month? The answer is 1, 1, 2, 3, 5, 8, 13 . . . – the Fibonacci numbers!

Ancient Greek mathematicians discovered that when you square every odd number above 1, you get a number that is a multiple of 8 plus 1. For example, $5^2 = 25$, which is $3 \times 8 + 1$.

Light takes 8 minutes to reach Earth from the sun.

In Islam there are 7 hells but 8 paradises, because God's mercy is greater than his wrath.

The average American uses 8 times as much fuel energy as the average person elsewhere in the world. The average man anywhere, just about, thinks about sex every 8 minutes.

The average person eating an average chocolate bar will not be pleased to hear that, on average, it will contain 8 insect legs. But that's nothing: the average human eats 8 spiders in their lifetime.

 There are 8 notes in the diatonic scale, which most western musicians use, some better than others. Compare Beethoven, say, with the Backstreet Boys, for example.

And by the way, $8 = 2^3$, or '2 cubed'.

There is a simple test to find out whether a number can be divided by 9: if 9 divides the sum of the number's digits, the number is divisible by 9. So, for 279, $2 + 7 + 9 = 18$, which is divisible by 9. And $279 = 31 \times 9$, so, yeah, it's divisible by 9.

Nine pops up all over the place. Troy was besieged for 9 years and Odysseus travelled for 9 years. In Ancient Mexican cultures there were 9 rivers in the lowest layer of the netherworld, while the River Styx in the netherworld of Ancient Greece had 9 twists.

China's netherworld also had 9 rivers, which were manifested in the 9-headed dragon, killed by the hero Yu, to whom the turtle then brought the 1st **magic square**, consisting of 9 numbers (see below), engraved on its back. In Chinese belief, the sky has 9 parts, the earth 9 countries, and each country 9 mountain ranges. The design of the Chinese capital, Beijing, consists of 8 streets leading to a centre (making 9).

The 1st 9 whole numbers, 1–9, can be arranged in a magic square so that all rows and both diagonals have the same sum: 15. This is the magic square brought on the turtle's back in Ancient Chinese myth, and it's called the *Lo Shu*.

There is really only 1 way to achieve the *Lo Shu* (there are variations, but they're essentially the same thing). What is it?

Genghis Khan was attended by 9 yaks' tails (his standards) and people had to prostrate themselves before him 9 times. To be honest with you, despite the possible effect on my dignity, I'd have prostrated myself 999 times if Genghis had asked – he was just that sort of guy.

A cockroach can live for up to 9 days without its head before it starves to death.

In the 9th episode, called 'Nasty', of the 12 episodes of the UK comedy *The Young Ones*, Neil has a bath.

Incidentally, $9 = 3^2$.

Our number system is **base 10**. This means that after 10 numbers in a position, we move to the next place. So 37 means 3 lots of 10 and 7 lots of 1. If we used, say base 7, then the number 25 would be $2 \times 7 + 5 \times 1$, which is 19 in our base 10. In base 6, 312 would be $3 \times 6 \times 6 + 1 \times 6 + 2$ which is 116 in base 10.

For Pythagoreans, 10 was the all-embracing, all-limiting 'mother' number. Ten is a significant number in many cultures: Aristotle recognised 10 categories, Moses read 10 commandments, and Buddhism, too, has 10 commandments.

When a Roman army revolted, every 10th soldier was executed (that's how you decimate your troops).

Seventies punk legends The Sex Pistols played for a full 10 minutes at their 1st gig in London on 6 November 1975. They fitted a lot of notes into those 10 minutes, though.

10cc was the name of a seventies UK 'art-pop' band. They called themselves after the average amount of semen a male ejaculates at orgasm: 10 cubic centimetres. Other bands named after semen include Pearl Jam and The Lovin' Spoonful.

The average life of a human tastebud is 10 days, and much less if it comes into contact with my cooking.

Check out this pattern: $1^2 = 1$, $11^2 = 121$, $111^2 = 12\,321$. Next time you really want to impress someone, point out that:

$$111\,111\,111^2 = 12\,345\,678\,987\,654\,321$$

 Eleven is the number of Tim Tams in a packet. As 11 is a prime number, this makes it a bugger to share a pack evenly among family or flatmates. Tim Tam lovers will be sorry (but probably not surprised) to hear that 11 was considered by Christians to be the number of sinners, while Muslims called it the 'mute number'.

In Ancient Sparta a group of 11 women was formed to stop the orgies of the Dionysian cult. We don't know just what they planned to do, but it is unlikely to have involved Tim Tams.

Eleven is the loudest the Marshall guitar amps go in the crocumentary *This is Spinal Tap* (the band in the movie, also called Spinal Tap, bore an uncanny resemblance to Black Sabbath). As the Tap's guitarist, Nigel, explained, 'Most blokes will be playing at 10. What we do is, if we need that extra push over the cliff [we go to] 11. One louder.'

As Nigel clearly knew, 11 is 1 more than 10.

Twelve is the 1st **abundant number**. A number is abundant if the sum of all its factors is greater than the number itself. So 12 can be written as 1 × 12, 2 × 6 or 3 × 4. Add up 1 + 2 + 3 + 4 + 6. They equal 16, and 16 is bigger than 12, so 12 is abundant.

QUIZ QUESTION

There are 5 other abundant numbers less than 40. What are they?

Twelve hours: that's how long snails take to have sex.

There were just 12 episodes of *Fawlty Towers* made. At the beginning of each episode, not unlike Bart writing on the blackboard in *The Simpsons*, the hotel sign was a variation of the words 'Fawlty Towers'. In the 12th episode, called 'Basil the Rat', the sign said 'Farty Towels'.

There are 12 notes in the chromatic scale of western music, which is all the notes between, say, A and the next A an octave higher.

Twelve is not a weird number. Believe me. Read ➤70 to find out why.

Thirteen is a Fibonacci number (see ➢3). The Fibonacci numbers regularly occur outside of mathematics, in nature. Almost all apples, roses, blackberries, raspberries, strawberries, peaches, plums, pears and cherries have flowers with 5 petals, and 5 is also a Fibonacci number. Pineapples have either 8 or 13 spirals coming out of their top. These are called **parastichies**.

If you freaked out when you saw this page, you may suffer from **triskaidekaphobia** – the fear of the number 13. We're so scared of the number 13 it costs the world billions of dollars a year in absenteeism, travel cancellations and loss of business. There were 13 at the Last Supper, witches apparently hang out in groups of 13, and, in China, the 13th month that was occasionally added to bring the lunar year into line with the solar year was called 'Lord of Distress' or 'Oppression'. In most casino hotels you won't find rooms numbered 713 or 2613, for instance, because they contain the number 13. In fact, often the whole 13th floor is left out, just to be safe.

If it's Friday while you're reading this, and it's the 13th, you'll know this month started on a Sunday. Makes sense when you think about it.

French king Louis XIII announced that 13 was his favourite number. Understandable, really. He married Anne of Austria when she was just 13 years old. Classy work, Louis.

There are 14 pounds in the imperial weight measurement, the stone. Fourteen is also the number of days in a fortnight. The Arabic alphabet consists of 14 sun letters and 14 moon letters, and one of Muhammad's names, *Taha*, has the numerical value of 14. And a garden pea has 14 chromosomes.

There are 14 vertebrae in the lower part of the spine and 14 in the upper, and 14 parts to the human hand.

To make a perpetual calendar you need 14 individual calendars.

Those French kings are at it again, as are the numerology enthusiasts: Louis XIV ascended the throne in 1643, the sum of whose figures add up to 14; he died in 1715 (again, the digits total 14), reigned for 77 years (14), was born in 1638, died in 1715 and $1638 + 1715 = 3353$ (14). All of which probably means – well, stuff all, to be honest. He didn't star in the 1951 movie *Fourteen Hours*. Good old Barbara Bel Geddes did, though (unless she is really Louis XIV), but that would be a bit freaky, even for those kooky French kings.

Curiously, $10^2 + 11^2 + 12^2 = 13^2 + 14^2$. Give it a go and convince yourself.

Fifteen is the 5th **triangular number**. Triangular numbers come from the number of dots in a right-angled triangle diagram like this one:

They are formed by adding up the series $1 + 2 + 3 + 4 + 5 \ldots$ Some triangular numbers are also square numbers (36, for example). But no triangular number can be a cube or a 4th or 5th power. To find out, say, the nth triangular number or T_n, use the formula $T_n = \frac{1}{2} \times n \, (n + 1)$. So:

$$T_5 = \frac{1}{2} \times 5(5 + 1)$$
$$= \frac{1}{2} \times 5 \times 6$$
$$= 15$$

QUIZ QUESTION

What are the 6th, 7th and 8th triangular numbers?

Triangular numbers can be **palindromic** as well. That is, they read the same forwards or backwards. But 15 isn't one of them.

Fifteen is incredibly important in western culture, being the number of coloured balls on a pool table: 7 big ones (stripes), 7 small ones (plain) and the 8 ball. It's important for another reason, too, being the number of minutes we can all expect to be world-famous, if Andy Warhol is right.

The famous Swiss mathematician Leonhard Euler proved that 16 is the only number that can be written in a **reverse power** way: $16 = 2^4 = 4^2$.

Sixteen was proposed as the basis for a new system of counting by J.W. Mystrom, who thought we could name the numbers in the new 16-base system an, de, ti, go, su, by, ra, me, ni, ko, hu, vy, la, po, fy and ton. Nice idea, J.W. – we'll get back to you.

Ancient Romans divided a foot (the measurement) into 16 fingers. See ➤4 if you're not convinced.

Sixteen was a favourite number in India and, until recently, the rupee was divided into 16 anna. In classical Indian aesthetics, there are 16 signs of beauty.

On his last 2 tours, Elvis wore only his resplendent sundial suit, which featured a sundial with 16 points on the back. The suit featured a lot more, as well, such as Elvis's own substantial body mass. Of course, Elvis recorded his 1st record at Sun Studios, in Memphis, Tennessee. And he recorded 16 tracks there. Elvis is now working at a truckstop on Highway 61, as well as a hairdresser's in Nashville and 1 in 10 Seven-Elevens worldwide.

Consider this: $17 = 2^4 + 1$. Now, $4 = 2^2$, so we can write $17 = 2^{2^2} + 1$. This makes 17 the 2nd **Fermat number**. Fermat numbers follow the formula $F_n = 2^{2^n} + 1$. So the 3rd Fermat number is $2^{2^3} + 1 = 2^8 + 1 = 257$. And $F_4 = 2^{16} + 1 = 65\,537$. In 1640, Pierre de Fermat claimed that all Fermat numbers were prime. But he turned out to be wrong. $F_5 = 2^{2^5} + 1 = 2^{32} + 1 = 4\,294\,967\,297 = 641 \times 6\,700\,417$. Bad luck, Pierre.

Seventeen is important in the Islamic tradition. The great Muslim alchemist Jabir ibn Hayyan believed that 17 was the basis for the material world, which consisted of the series 1, 3, 5 and 8. The cycles of prayer movements in the 5 daily prayers amounts to 17 and 17 is the number of words in the call to prayer.

Big Arnie Schwarzenegger really had to stretch himself in *Terminator*, in which he speaks a massive 17 sentences in his own voice. Two other lines coming out of his mouth are overdubbed with the voices of other characters (because he has taken on their forms).

Eighteen is the 6th **Lucas number**. The Lucas numbers start with 1 and 3, then, following the same pattern as the Fibonacci numbers, we add the last 2 numbers to get the next one. So, 1, 3, 4, 7, 11, 18 . . . Eighteen is also an abundant number.

If there are 18 or more people at a party, then there must be either a group of 4 people who all know each other or a group of 4 complete strangers.

If you want to become a Whirling Dervish, that is, a member of the Mevlevi, you'll need to serve for 18 days in the monastery as a houseboy, then learn the 18 different kinds of service in the kitchen. Once you've completed the 1001 days of preparation, you'll be led with an 18-armed candelabrum to your new cell and you'll meditate for 18 days. Then, and only then, can you start whirling. I don't know anyone who's done it, but I'm not willing to write it off.

Eclipses of the sun and moon recur in the same sequence after 18 years.

The cuddly, fluffy cute little individual that pads around the house has 18 claws. Of course I mean your cat.

Remember the test to see if a number is divisible by 3 (see ➤3)? Well, there's also a simple test to see if a number is divisible by 19. If $100a + b$ is divisible by 19, then $a + 4b$ must be divisible by 19. Let's test 323:

- $323 = 3 \times 100 + 23$, so $a = 3$ and $b = 23$.
- Therefore, $a + 4b = 3 + 4 \times 23 = 95$.
- And 95 is divisible by 19. So 323 is divisible by 19 (it is: it's 19×17).

QUIZ QUESTION

Try the same test to see which of these 4 numbers is divisible by 19: 209, 363, 418, 817.

If you're an Ancient Babylonian (lots of people were), you may well suffer from **nonadekaphobia**, because the 19th day of the month, being the 49th day from the beginning of the previous month (or 7×7) was considered to be filled with power, both good and evil.

The Metonic cycle of the moon occurs every 19 years, when all the phases of the moon fall on the same days of the week during the entire solar year.

The Japanese board game Go is played on a board 19 squares by 19 squares.

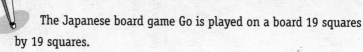 Nineteen is a prime number.

In case you hadn't noticed, fingers and toes add up to 20, and so 20 was the basis for counting in many cultures (see ➤10 if you're not sure what I'm going on about). The Ancient Mayans of Central America used 20 as the basis for their number system. As well, their calendar was made up of 13 months of 20 days each, and in their writing system the symbol for the number 20 was a human figure, probably because humans had 20 fingers and toes.

Twenty is an abundant number:

$$1 + 2 + 4 + 5 + 10 = 22.$$

The biggest **icosahedron** (20-sided figure) is in Lexington, Massachusetts.

Soccer teams have 11 players, hence a standard game sees 22 players on the field. Now, if 2 people are sent off and only 20 remain, that's usually considered a fairly spiteful game. So what does that say about a game played in Paraguay between Sportivo Ameliano and General Caballero, where a record 20 players were sent off and only 2 remained?

Jerry Seinfeld, in his TV show, took 20 years to read Henry Miller's *Tropic of Cancer*, which he borrowed from the New York Public Library and failed to return.

20/20 is the name of a Beach Boys album. But then again, so is *Pet Sounds*.

Twenty-one is the 8th Fibonacci number. It is also the 6th triangular number. Using the formula from ➤15:

$$T_6 = \tfrac{1}{2} \times 6(6 + 1) = 21.$$

 Twenty-one is the number of snooker balls, not counting the cue ball (the white one).

When they turn 21, Americans can have a beer to celebrate the fact that they've been eligible for 3 years to vote and to be shot in a war. Lucky, eh? In many countries it is virtually illegal to turn 21 without enduring a public humiliation that includes drinking from a ridiculously long glass, having any and every low point in your love life recounted by 'friends', and putting up with inebriated uncles who can't dance but will try to hit on women 21 years their junior.

The original *Halloween* movie was shot in 21 days in 1978, on a budget of $300 000. This made the story of the murderous Michael Myers the highest-grossing independent movie ever made. That title has now been taken by *The Blair Witch Project*. The film itself isn't that scary, but it's absolutely terrifying that so many people parted with good money to see it.

Elephants are pregnant for 21 months.

Probably the most famous number of all is π – pi – the ratio of the circumference of a circle to its diameter. In real words, this means that if a circle is 3 centimetres across, then it's 3 × π centimetres around the outside.

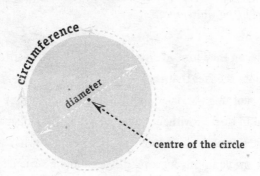

circumference

diameter

centre of the circle

$$\pi = \frac{\text{length of circumference}}{\text{length of diameter}}$$

But π is an irrational number, so the best we can do is make an approximation of it. As a decimal, π equals roughly 3.14159. And this is where 22 comes in: the most commonly used fraction approximation is $\pi \approx \frac{22}{7}$. So if a circle is 7 cm across, it's roughly 22 cm around.

The square of 22 is a palindrome (it reads the same forwards or backwards): $22^2 = 484$.

QUIZ QUESTION

There are 5 other numbers less than 100 that have palindromic squares. One, 2 and 3 are obvious, because the squares have only 1 digit. Find the other 2 (they're both below 30).

The saying 'It's a catch-22' comes from Joseph Heller's famous book about World War II called, you guessed it, *Catch-22*.

Remember triangular numbers (see ➤15)? Well, after triangular numbers and squares come **pentagonal numbers**, from the 5-sided figure the pentagon. The idea is the same: we keep expanding by putting another pentagon on the outside. The pentagonal numbers are the total number of points in each stage of the diagram. Here's 22:

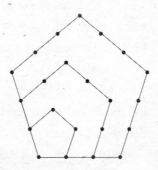

QUIZ QUESTION

Can you get the next 3 pentagonal numbers after 1, 5, 12 and 22?

The strange-looking 23! is 23 digits long. See ➤24 to find out what the hell this means.

Earth is tilted to about 23 degrees. See ➤98 for more on planetary tilts.

In a recent poll, when asked what they preferred to do in bed other than sleep, 23 per cent of women said they'd prefer to read – the most popular response among female respondents. Of the men polled, 50 per cent said they'd prefer to have sex. Only 20 per cent of women responded the same, which leaves 30 per cent of blokes to amuse themselves.

If there are 23 or more people in a room, the odds of 2 people sharing a birthday are better than 50:50. For fewer than 23, it's more likely that nobody will share a birthday. This is a surprisingly small number, but it's true. Once you get above 40 people, it's very likely, and a group of 45 people have a staggering 95 per cent chance of a shared birthday. I'll show you that 23 is the break-even point for the shared-birthday party trick. In fact, as is often easier in probability questions, we'll work out the odds of the exact opposite happening.

- Take 1 person. They have a birthday on some day – say 1 January. Choose a 2nd person. The odds that their birthday does not clash with the 1st person's is $\frac{364}{365}$, because there are 364 days apart from 1 January.

- Choose a 3rd person. The odds that their birthday falls on a separate day again is $\frac{363}{365}$, because we want their birthday to be different to the 1st two. So, in a group of 3 people, the chance of the 2nd person not clashing birthdays with the 1st, and the 3rd person not clashing with either of the others, is $\frac{364}{365} \times \frac{363}{365}$.

- For a group of 4, we follow the same logic and the odds are $\frac{364}{365} \times \frac{363}{365} \times \frac{362}{365}$ that there is no common birthday. Can you see the pattern?

- For larger groups, we just keep multiplying terms on the end. So if you have a group of 23 people, the odds of no shared birthdays are $\frac{364}{365} \times \frac{363}{365} \times \frac{362}{365} \times \ldots \times \frac{343}{365}$, which is about 0.49. So there is a 51 per cent chance of a shared birthday!

According to records, the longest that a red kangaroo has bounded about is 23 years, even though the average life span for a red kangaroo is 4–6 years.

Because 24 = 4 × 3 × 2 × 1, we call it 4 factorial or 4! **Factorials** occur all the time in mathematics, especially in probability and counting problems. Say I had 3 different books and I put them in a pile. There are 3 possible books that could go on the bottom, then 2 choices for the middle, and 1 book left to go on top. So there are 3 × 2 × 1 = 3! = 6 ways of making the pile. By the same thinking, 4 books can be stacked 4 × 3 × 2 × 1 = 24 different ways.

From alpha to omega and aleph to tof, there are 24 letters in both the Greek and the Hebrew alphabets. Twenty-four is also the number of hours in a day. If you're lucky enough to be in a certain religious theme park in Las Vegas you can catch the Red Sea parting and Lazarus rising from the dead every single one of those 24 hours.

It was the Babylonians who divided the day into 24 hours and then divided each hour into 60 minutes and each minute into 60 seconds. This system has survived for 4000 years.

The Bee Gees recorded a number of songs for *Saturday Night Fever*, including the opening theme, 'Stayin' Alive'. When the film premiered in 1977, not only did it establish the Hollywood career of star John Travolta, it created a national dance craze and influenced fashions for the 1970s. So where does 24 fit into this? Well, the soundtrack topped the US charts for 24 weeks.

Twenty-five is a **lucky number**. No. I'm not talking some bizarre numerology-aromatherapy gooshta here. To find the lucky numbers, write out a long list of numbers, say 1–100. Now look at number 2, the 2nd number on the list, and remove it and every 2nd number after it. The list now reads 1, 3, 5, 7, 9 . . . The next number on the list is 3, so now, remove every 3rd number. The list is now 1, 3, 7 . . . Remove every 7th number still on the list. Now remove every 9th number on the remaining list. Continue this pattern, or 'sieve', as mathematicians call it, and you're left with the lucky numbers.

QUIZ QUESTION

What are the next 5 lucky numbers after 25?

Twenty-five is also a perfect square: $5^2 = 25$. Now, the product of any 4 consecutive numbers + 1 will always equal a square. So:

$$1 \times 2 \times 3 \times 4 + 1 = 25 = 5^2$$

or

$$11 \times 12 \times 13 \times 14 + 1 = 24\,025 = 155^2$$

For those of you who can multiply algebra, the proof isn't too hard to follow:

$$n(n + 1)(n + 2)(n + 3) + 1$$

which is just 4 consecutive numbers plus 1, will open up into

$$n^4 + 6n^3 + 11n^2 + 6n + 1$$

which is just $(n^2 + 3n + 1)^2$.

If there are 25 or more people at a party, then there must be either a group of 4 people who all know each other or a group of 5 complete strangers.

If you're turning 25 this year, it's your silver jubilee. See. Told you it was lucky.

Twenty-six is the smallest number that isn't a palindrome but has a square which is: $26^2 = 676$. Amazing, hey?

There are 26 characters in the alphabet used in English, which is called the modern Roman alphabet. There used to be only 25, until J came along in the 14th century.

The great English Romantic poet, John Keats, died at the age of 26. So did the English artist Aubrey Beardsley. And the brilliant Norwegian mathematician Niels Abel. They all died of tuberculosis, but to suggest any link here would be real conspiracy theory stuff.

The number 10^{27}, or 1 with 27 zeros after it, is called an octillion. In case you're wondering what one looks like, here's one I prepared earlier:

1 000 000 000 000 000 000 000 000 000

Twenty-seven is the smallest number that can be written as the sum of 3 squares in 2 different ways.

QUIZ QUESTION

Try to find the 2 ways you can write 27 as the sum of 3 squares.

I've already mentioned what a major gun the Swiss mathematician Leonhard Euler was (see ➤16), but, hey, sometimes even he screwed up. Consider this:

- Triangles in high school give us the equations $3^2 + 4^2 = 5^2$ and $7^2 + 24^2 = 25^2$ etc.

- It's also true that $3^3 + 4^3 + 5^3 = 6^3$.

- But we can't get 2 cubes to add up to a cube. There are no whole numbers a, b and c that solve $a^3 + b^3 = c^3$ (see ➤67 for a longer explanation).

- Euler saw all this going down and suggested that $a^4 + b^4 + c^4 = d^4$ and $a^5 + b^5 + c^5 + d^5 = e^5$ etc. would also have no solutions, or, to use mathspeak, 'You can't write an nth power as the sum of $n - 1$ nth powers.' Pretty groovy-sounding, hey!

- But in 1966, 2 mathematicians called Leon Lander and Thomis Parkin showed that

$$27^5 + 84^5 + 110^5 + 133^5 = 144^5$$

thus torpedoing Leonhard's really cool conjecture. That's the way it goes, big guy. Even the best of us etc., etc. . . .

Kurt Cobain died at the age of 27. So did Jimi Hendrix. And Janis Joplin. And Jim Morrison.

Oh, and just in case you didn't realise, $27 = 3^3$.

Twenty-eight is the 2nd perfect number after 6 (if you've forgotten what a perfect number is, go back and take a look at ➤6) and the 7th triangular number (see ➤15 if you've forgotten that as well).

Twenty-eight is also the 12th **Ulam number**. Now, you're probably wondering where you can go to find out what that is. Take a look at ➤47.

Twenty-eight is the number of days in the lunar cycle, the number of days in a textbook menstrual period, the number of dominoes in a standard set and the number of properties on a Monopoly board.

Seven cuts through a pizza can create up to 29 pieces. Try it yourself with a pencil and a big circle drawn with a pen. But next time you've got 29 friends coming over, don't think 'Hey, I only need 1 pizza', because it's a bugger to work out and some of the pieces will be pretty small. In general, to find out the maximum number of pieces you can get from a certain number of cuts, think of the number of cuts as n and put n into the formula:

$$\text{pieces of pizza} = \frac{n \times (n + 1)}{2} + 1$$

So with 13 cuts, you could make 92 pieces of pizza:

$$\frac{13 \times (13 + 1)}{2} + 1 = \frac{13 \times 14}{2} + 1 = 13 \times 7 + 1 = 92.$$

Please note that if you did manage to do this, several of your friends would be *very* dirty about the size and shape of their piece.

QUIZ QUESTION

If you've got 56 friends over, what's the smallest number of cuts you could make to your pizza and get 56 pieces?

The longest word in the *Oxford English Dictionary* has 29 letters. **Floccinaucinihilipilification** means 'to estimate as worthless'. To use it in everyday conversation, try this.

You: Hey, Sarah. What do you think of Celine Dion's music?

Sarah: I think it's absolute crap!

You: Well, now, there's no need for floccinaucinihilipilification.

One year on Saturn equals 29 Earth years. So New Year's Eve on Saturn tends to get pretty crazy.

Thirty is a **primorial**: 30 = 5#. I mentioned factorials earlier (see ➤24). To get a factorial you multiply a number by all the whole numbers less than but including itself. So 5 factorial or 5! = 5 × 4 × 3 × 2 × 1 = 120. Now, a primorial is where you multiply a prime number by all the prime numbers less than itself. So 5 primorial, written 5#, is 5 × 3 × 2 = 30.

Thirty is connected with order and justice: in Ancient Rome, a man had to reach the age of 30 in order to become a tribune, and both Moses and Jesus began to preach at that age. In modern times it is the age at which you realise you are no longer in your early twenties.

A lunar month, the amount of time that it takes the moon to pass through its phases and orbit the Earth, is close to 30 days – more like 29.5306, if you're fussy – which is why many ancient calendars agreed on 30 days as the length of a month. The problem with this was that even the ancients knew that it took about 365 days for the sun to return to rising in exactly the same position as it did on a given day, for the seasons to repeat, and so on. So the solar year, which is how long it takes the earth to orbit the sun – though most of the ancients didn't realise that the earth went around the sun (but that's another story altogether) – wasn't an exact number of 30-day months. As a result, some extra days needed to be added. There's a great book by David Ewing Duncan called *The Calendar* which goes through the whole saga. See ➤46 for some more wacky calendar antics.

There is enough Lego on the planet for every person to have 30 pieces each.

Because $31 = 2^5 - 1$, it is the 3rd Mersenne prime. Almost all of the largest prime numbers ever discovered are Mersenne primes ($2^p - 1$, where p is prime). The 37th Mersenne prime, discovered in 1998, is $2^{3\,021\,377} - 1$. It has just 909 526 digits. But we know that this gigantic number is prime. Mind-blowing, hey? On 1 June 1999, a new record prime was discovered by Nayan Hajratwala: $2^{6\,972\,593} - 1$. This is the 38th *known* Mersenne prime (there may be smaller ones, because not all the numbers between the 37th Mersenne prime and this one have yet been checked). It has 2 098 960 digits, and if typed out in this format it'd run to over 2915 pages. To find out more about these bizarre numbers, search the Web for Mersenne primes or for GIMPS, the Great Internet Mersenne Prime Search, which is offering up to US$250 000 to the first home-computer user – maybe you – who discovers a 1 billion digit Mersenne prime.

Leeches have 31 brains more than a human, that is, they have 32. Despite this, leeches have been responsible for absolutely none of the mathematical milestones in this book.

The very 1st crossword appeared in 1913 in the Sunday attachment to the *New York World*. It was written by Arthur Wynne, an Englishman, and contained 32 clues.

Thirty-two degrees is the point at which water freezes on the Fahrenheit scale. Why? Because the 18th-century German physicist Daniel Fahrenheit chose the temperature of an equal ice–salt mixture as zero for his scale. He identified the freezing point of water as 30 on his scale, but this was later revised to 32.

Bart Simpson is *not* a 32-year-old woman, if you can believe anything he writes on the blackboard at the start of *The Simpsons*.

Scott, Virgil and Alan jangled their way through 32 episodes of *Thunderbirds*.

And $32 = 2^5$.

Here's a bizarre-looking equation:

$$33 = 1! + 2! + 3! + 4!$$

See ➤24 if you don't know what I'm going on about.

 Would you believe that there is a statue of Popeye the sailor man in Crystal City, Texas, proudly commemorating the fact that the fantastically forearmed freak singlehandedly increased their consumption of spinach by 33 per cent?

Ronald Belford Scott (or 'Bon' to his mates), the legendary singer of those dirty deeders, AC/DC (the name came from their mum's vacuum cleaner), was just 33 when he died. Strangely, he looked a fair bit older.

Jesus Christ also died when he was 33. So did Blues Brother John Belushi. The song played at Belushi's funeral was the Ventures' 'The 2000-Pound Bee', which cannot be found in the Book of Psalms.

Thirty-three is the 10th lucky number.

In a 4 × 4 magic square, the numbers 1–16 occur once each and all the rows, columns and diagonals add up to 34.

QUIZ QUESTION

Try finishing off these 4 × 4 magic squares:

12	13		8
6			10
			11
9			5

	2		16
13		3	
	7		5
8		6	

10	16	1	7
15	5	12	2

3	15		14
13			
			1
		7	

What's a lummox? Well, it's a clumsy, stupid person. What's so special about a lummox called Jiminy? Well, he is in fact Stimpy's conscience, whom Stimpy gives to Ren for a day. Jiminy likes to beat up on Ren for every misdeed. And what's the significance of all this? Well, it all takes place in the 34th episode of *Ren and Stimpy*, which is called 'Don't Drag Your Butt Over This Document'.

A polyomino (not as scary as it sounds) with 6 squares is called a **hexomino**. A fun way to spend an afternoon is to try and find all 35 hexominoes. Here are 4 to get you started:

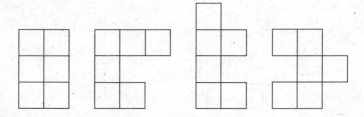

Thirty-five is the number of seconds that the shortest play ever recorded runs to. It's Samuel Beckett's *Breath*, and it consists of 35 seconds of human cries and breaths. A tough one for the critics to review, I'd have thought, and very embarrassing to turn up to late.

In chess, a knight moves 2 squares forward or to the side, then 1 square around the corner. So the knight on the middle square could move to any of these 8 positions:

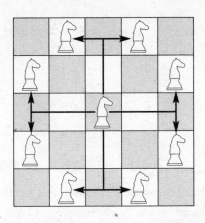

On a board 8 squares long and 8 wide, a knight can travel 35 steps without crossing back over its path. See if you can find the 35-step path.

If you're a crocodile with a bunch of eggs on the way and you're wondering what sex they'll be, check the air temperature. Crocodile eggs that are incubated above 35°C (95°F) hatch into males; below 29.5°C (85°F) they hatch into females.

If you add all the numbers between and including 1 and 36, which are all the numbers on a roulette wheel, you get 666, a popular number with the devil.

Thirty-six is the 1st number that can be written as the sum of 3 separate cubes: $36 = 1^3 + 2^3 + 3^3$. The other 3 numbers under 100 that can be written this way are:

$$71 = 1^3 + 2^3 + 4^3$$
$$92 = 1^3 + 3^3 + 4^3$$
$$99 = 2^3 + 3^3 + 4^3$$

Homer Simpson is 36 years old. See also ➤81.

Thirty-six is what's known as a **highly composite number**. A number is composite if it has divisors other than 1 and itself. So 14 is composite because $14 = 2 \times 7$, but you can easily check (by a bit of division) that 17 can only be written as 17×1. Seventeen isn't composite; it's a prime number. Thirty-six, on the other hand, is not only composite, it's highly composite. This was a concept invented by the brilliant Indian mathematician Ramanujan for numbers that have more divisors than any other below them. So 36 has divisors 1, 2, 3, 4, 6, 9, 12, 18 and 36. No number from 1 to 35 has 9 divisors. Other highly composite numbers include 332 640, 43 243 200 and 2 248 776 129 600.

QUIZ QUESTION

Start from 1 and work out the highly composite numbers before 36.

 Thirty-six is the 8th triangular number, and an abundant number (see ➤12 if you don't believe me).

According to *Billboard* magazine research, Kiss is currently the 36th biggest-selling album act in the rock era, placing them ahead of such acts as Pink Floyd, The Grateful Dead, Led Zeppelin, Billy Joel, The Who, Aerosmith and Van Halen. So throw that in Dad's face the next time he tries 'No son of mine's wearing make-up to a party!'

And finally, $36 = 6^2$.

Thirty-seven is the 4th **centred hexagonal number**. We get these by placing hexagons around a centre. Not surprising really. So the centred hexagonal numbers are 1, 7, 19 etc.

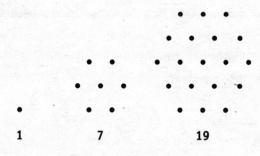

1   7   19

QUIZ QUESTION

What are the next 3 centred hexagonal numbers after 19?

There's a cute fact about the number 37 and when it divides into other numbers. If 37 divides a number abc, then it divides the number bca. For example, 37 divides into 259 ($37 \times 7 = 259$), so it divides into 925 ($37 \times 25 = 925$).

If you're feeling hungry, 8 cuts to a pizza can create up to 37 pieces. Go get it!

You can find the number 37 at the 4 corners of the world. Now, the expression 'the 4 corners of the world' generally means the remotest parts of the planet. But the earth has no corners, in the technical sense. The notion that it does probably derives from the time when the earth was thought to be flat. However, the earth is not a perfect sphere, and a few places 'bulge' a bit. With this geological phenomenon in mind, the 4 corners of the earth were identified in 1965: they are Ireland, south-east of the Cape of Good Hope, west of the Peruvian coast, and between New Guinea and Japan. Each of these 'corners' is several thousand square kilometres in area and 37 metres above the geodetic mean, and the gravitational pull is measurably greater at these locations.

There are currently thought to be 37 plays written by Shakespeare. If you try to write out a list, don't forget that *Henry IV* has 3 parts and that *Hamlet* was written by Shakespeare, not Mel Gibson.

Since 1978, at least 37 people have died as a result of shaking vending machines in an attempt to get free merchandise. More than 100 have been injured. I'm not sure exactly how sorry I feel for these people.

Remember magic squares? (Check out ➤9 if you've forgotten.) Well, there is only 1 magic hexagon. It uses the numbers 1 through to 19, and each side and each diagonal adds up to 38.

QUIZ QUESTION

Try and finish the magic hexagon.

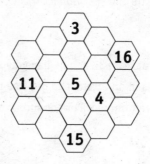

The shortest war ever recorded was between Britain and Zanzibar in 1896. It lasted just 38 minutes, but would still make a great Tom Hanks film.

If you want to be as cool as Sean Connery's James Bond (don't we all?), never drink Dom Perignon '53 above the temperature of 38°F. It's just not done, according to 007 in *Goldfinger*.

Spot the continuity error: in *Raiders of the Lost Ark* during the firefight in the bar, our hero Indiana Jones starts out with a .38 revolver – which changes to a .45, back to a .38, and back once again to a .45.

Thirty-nine has been described as 'the 1st **uninteresting number**'. While it's not as sexy as 7 or as nifty as 99, it's got a few things going for it. For example: $39 = 3 \times 9 + 3 + 9$. Not impressed? Well, this pattern holds for any number ending in 9. Why? It's simple.

- A number with digits ab is equal to $10a + b$. So here, $39 = 3 \times 10 + 9$.
- So the example $39 = 3 \times 9 + 3 + 9$ matches up $10a + b$ and $a \times b + a + b$. Let's solve that $10a + b = a \times b + a + b$.
- Subtract b from each side and you get $10a = a \times b + a$.
- Subtract a from each side and you get $9a = a \times b$.
- As long as a isn't a zero, divide each side by a and get $b = 9$. So:

$$39 = 3 \times 9 + 3 + 9$$
$$79 = 7 \times 9 + 7 + 9$$
$$159 = 15 \times 9 + 15 + 9 \text{ etc.}$$

Most high-school mathematicians remember π as 3.14 or $\frac{22}{7}$ (see ➤22). In fact, it has been calculated to over 50 billion decimal places, and individuals have recited it from memory to over 4000 places. While this is bizarrely impressive, it's quite unnecessary. If you used π to 39 decimal places to calculate a circle surrounding the entire known universe, you would be accurate to within the width of 1 hydrogen atom. Of course, that approximation for π would be: 3.14159 26535 89793 23846 26433 83279 50288 4197 . . .

The number 10^{39}, or 1 with 39 zeros after it, is called a **duodecillion.** If you suspect there may be one hanging around your house, it looks a bit like this: 1 000 000 000 000 000 000 000 000 000 000 000 000 000.

In the 39th episode of *South Park*, called 'Two Guys Naked in a Hot Tub', Kenny doesn't die. That's because he doesn't appear. You bastards.

What are the 4 weights that can be used on a scale pan to weigh any whole number of grams from 1 to 40 inclusive, if the weights can be placed in either side of the scale pans?

This is the famous Bachet's problem from 1612. The answer is 1, 3, 9 and 27. To work it out, you have to notice that a scale can weigh something directly.

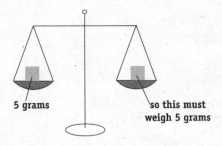

5 grams   so this must weigh 5 grams

But you can also weigh something by putting weights on both sides:

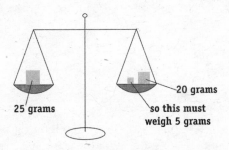

20 grams

25 grams   so this must weigh 5 grams

Convince yourself that you can use the 4 weights 1, 3, 9 and 27 to weigh everything from 1 gram to 40 grams.

Forty is an abundant number.

Forty is a popular number in the Bible: Moses spent 40 days and 40 nights on Mount Sinai before receiving the 10 Commandments; the Israelites wandered the desert for 40 years; Jesus fasted in the desert for 40 days and 40 nights and appeared to his disciples for 40 days after the resurrection.

Forty is the record number of years ever lived by a snake. It was a boa constrictor.

You only ever will see a rainbow in the morning or late afternoon, when the sun is 40 degrees or less above the horizon.

Minus 40 degrees Celsius is exactly minus 40 degrees Fahrenheit. It's the only time the 2 scales correspond.

The average cow produces 40 glasses of milk a day.

The famous Swiss mathematician Leonhard Euler discovered literally thousands of things in his incredible lifetime. One of them was the formula $n^2 + n + 41$, which looks harmless but has the amazing property that, for whole numbers $n = 0$, $n = 1$, $n = 2$ up to $n = 39$, it gives a prime number as its value. What this means is:

$$0^2 + 0 + 41 = 0 + 0 + 41 = 41 \text{ is prime;}$$
$$1^2 + 1 + 41 = 1 + 1 + 41 = 43 \text{ is prime;}$$
$$2^2 + 2 + 41 = 4 + 2 + 41 = 47 \text{ is prime;}$$
$$3^2 + 3 + 41 = 9 + 3 + 41 = 53 \text{ is prime;}$$

all the way up to

$$39^2 + 39 + 41 = 1521 + 39 + 41 = 1601$$

which is prime. However, $40^2 + 40 + 41 = 1681$, which is 41×41. Damn good effort, Leonhard.

Jane Austen died at the age of 41. She wrote 6 novels during that time (1775–1817), which is also a damn good effort and will keep Gwyneth Paltrow in work for years.

Everyone's heard the various theories about oldest brothers, middle sisters, etc. Well, I'm sure child psychologists around the world would leap at the chance to talk to the 2 children of Englishwoman Elizabeth Buttle, who were born a record 41 years apart.

Wolfgang Amadeus Mozart churned out 41 symphonies, the last of which was the Jupiter Symphony in C major. He wrote no film soundtracks and was never in a teenage boy band – though, despite being unaware of it at the time, he wrote hundreds of car commercials.

Any whole number can be partitioned. To **partition** a number is to write a whole number as the sum of other whole numbers. So $7 = 4 + 2 + 1$, $7 = 5 + 2$ and $7 = 3 + 1 + 1 + 1 + 1$ are all partitions of 7.

QUIZ QUESTION

Try and find all 42 ways of partitioning the number 10. Be careful not to repeat any, so set your calculations out in a sensible order.

In Japanese, the Arabic numerals 42 translate as *shi ni*, which is very close to the Japanese word *shinu*, which means 'to die'. As a result, 42 is a very superstitious number for some Japanese. Not Masao Asano, though, who lined up for the 1963 Grand Prix at the Suzuki circuit in Nagoya with his white Austin-Healey sporting the number 42. It sported 42 as it stood revving exquisitely on the starting line. It sported 42 as it raced towards the end of the 1st lap in the lead. Unfortunately for Masao Asano, it also sported the number 42 as it sped out of control, off the track and into a ditch, killing him. The Japanese Auto Federation subsequently banned the use of the number 42 on any vehicle competing in Japan.

According to Deep Thought, the computer in Douglas Adams's *Hitchhiker's Guide to the Galaxy*, the answer to Life, the Universe and Everything is 42. No-one really knows if this is true. But if it is, would the meaning of life become 46.2 under a 10 per cent consumption tax?

It takes 42 muscles to frown – but sometimes it's worth it.

Forty-three is the 12th lucky number. It's also a prime number.

QUIZ QUESTION

Including 43, there are 25 prime numbers between 1 and 100. Find them. (Note: 1 is not a prime).

As we go to press, the 43rd bestselling novel of all time is the paperback edition of Anne Rice's *Interview with the Vampire*, with around 3 million copies sold. The bestselling bestseller ever is the paperback edition of John Grisham's *The Pelican Brief*, with over 11 million copies sold. Obviously, *Adam Spencer's Book of Numbers* will change all that.

If you have 6 objects in a row, there are 44 ways of rearranging those objects so that none are in their original spot. Or, to put it another way, if 6 people are wearing different hats, and they all swap hats so none are wearing the hat they had been wearing, there are 44 different ways they could swap hats.

Drinking 1 cup of tea a day is thought to reduce your risk of major heart attack by up to 44 per cent.

Clint Eastwood's Dirty Harry carried a .44 magnum. The 1st movie in the series, *Dirty Harry*, appeared in 1971.

Though we have in our noses an impressive 5 million olfactory cells with which to smell, sheepdogs have 220 million, so they can smell 44 times better than we can. However, only some of us smell 44 times nicer than they do.

In 44 BC, just 2 years after he sorted out the calendar once and for all (see ➤46), Julius Caesar was assassinated. Bad luck, Jules, but at least we all agree on the date that it happened.

Forty-five is the 3rd **Kaprekar number**. When you square a Kaprekar number and add the n digit number on the right to the n or $n - 1$ digit number on the left, you get the original number. So:

$$45^2 = 2025 \text{ and } 20 + 25 = 45$$

QUIZ QUESTION

Which of the following 8 numbers are Kaprekar numbers?
51, 55, 59, 91, 99, 103, 295, 297

The body of the average adult human contains 45 litres of water, and as much carbon as a 12-kilogram bag of coke (the black stuff), 2200 matchheads worth of phosphorus, enough iron to make a 25-millimetre nail, and enough lime to whitewash a small shed. Note, none of these are easy, painless or legal to extract from your average adult human.

One of the highlights of Martin Scorsese's 1995 film *Casino* is Sharon Stone's performance as Ginger, wife of a mobster casino boss played by Robert De Niro. The gold and white beaded dress she wears in one scene weighed 45 pounds (20 kilograms), and would make the kinky leg moves she made famous in *Basic Instinct* an all-over body workout.

Nine cuts through a pizza can create up to 46 pieces.

How long is a year? I know how stupid that question sounds, but it's worth asking. You probably answered '365 days' or '365 or 366 days, depending on whether it's a leap year'. Perhaps you even know that the solar year (the time taken for the earth to fully orbit the sun) is roughly 365 days, 5 hours and 49 minutes – and I stress that's only *roughly*. But it hasn't always been that accurate. Around 700 BC, the Romans were using a 10-month, 355-day calendar. But it was obvious to the farming community that this was about 10 days short of the mark, because the seasons soon fell out of sync with the weather. To overcome this, the Roman priests and aristocrats tried various techniques based around inserting extra months now and then to top up the days that had been lost. The system didn't work smoothly due to a combination of factors: people were often confused about when the extra months were due, it was often politically expedient to screw with the process and change the dates that ceremonies and feasts were due, and the big cheese himself often had more pressing matters to address – I mean, if you had to choose between spending a few hours snogging Cleopatra or deciding whether February should be a few days longer, which way would you go? By 46 BC, the situation was way out of hand, and the calendar had slipped from the solar year by almost 2 months. Caesar rectified the situation in one fell swoop – he added the extra 23 days already due for February and another 2 months, of 33 and 34 days, between November and December. As a result, 46 BC, which Caesar called 'ultimus annus confusionus' (the last year of confusion) but most Romans just called the Year of Confusion, lasted a whopping 445 days.

The longest sausage ever recorded was 46 kilometres long. It was made by M & M Meat Shops and S.M. Schneider in Kitchener, Ontario, in 1995.

Currently 46 per cent of Australians are buried (see ➤54 to find out about the rest).

Forty-seven is a **Ulam number**. We get the Ulam numbers by writing 1, 2, 3 and then any other numbers that can be written *only once* as the sum of 2 other numbers on the list. So 1 + 3 = 4 and we put 4 on the list, but once we have 1, 2, 3, 4 we don't include 5 because 5 = 1 + 4 *and* 5 = 2 + 3. The list of Ulam numbers begins 1, 2, 3, 4, 6, 8, 11, 13, 16, 18 . . .

QUIZ QUESTION

Try to work out the Ulam numbers up to 100. It's bloody hard. Start with 19, and see if it can be written as only 1 sum from numbers on the list above. If so, put it on the list. If not, kick it. Then check 20, 21 etc.

What's a moop? In the 47th episode of *Seinfeld*, George gets into a fight with a boy who lives in a bubble over the correct answer to a Trivial Pursuit question. The bubble boy says the answer is 'the Moors' whereas George points out that the answer printed on the card is 'the Moops'. Obviously you can't always believe trivia you read.

In the great Billy Wilder comedy *Some Like It Hot*, Marilyn Monroe needed 47 takes to get 'It's me, Sugar' correct, saying instead either 'Sugar, it's me' or 'It's Sugar, me'. In another scene she had to rummage through some drawers and say 'Where's the bourbon?' Not too demanding, you would have thought. But after 40 takes of saying 'Where's the whiskey?', 'Where's the bottle?' or 'Where's the bonbon?', the director pasted the correct line in each drawer. She finally got it after 59 takes.

Forty-eight and 75 are **betrothed**. This happy couple have a relationship because the factors of 48 are 1, 2, 3, 4, 6, 8, 12, 16, 24 and 48. The sum of the factors of 48, except for 1 and itself, is 2 + 3 + 4 + 6 + 8 + 12 + 16 + 24 = 75. The sum of the factors of 75, excepting 1 and itself is, you guessed it, 48. Check that now. So 48 and 75 are betrothed. How cute!

QUIZ QUESTION

Which 2 of the following pairs are also betrothed? 84 and 88; 140 and 195; 108 and 171; 1575 and 1648.

A potato has 48 chromosomes, 2 more than most human beings. Which just goes to show that chromosomes are a start, but they're not everything.

Piglets are fast-growing little creatures. So much so that 48 of them were used during the filming of *Babe*.

Forty-eight is an abundant number.

One of Charlie Sheen's more memorable roles was as a drug addict at the police station in *Ferris Bueller's Day Off*, in preparing for which he did not sleep for 2 days. Charlie Sheen has stayed awake for more than 48 consecutive hours many times in his life, but this is the only known instance of him doing it for artistic reasons.

Forty-nine is the 13th lucky number. It's also a perfect square: $7 \times 7 = 49$. Now, watch, learn, then impress your friends:

$$7 \times 7 = 49$$
$$67 \times 67 = 4489$$
$$667 \times 667 = 444\,889$$
$$6667 \times 6667 = 44\,448\,889$$

QUIZ QUESTION

What's 666 667²?

The longest that a hippopotamus has wallowed in mud and generally had a fine time is 49 years. The average life span for a hippo is 30 years.

Fifty is the smallest number that is also the sum of 2 squares in 2 different ways. So $5^2 + 5^2 = 50$ and $7^2 + 1^2 = 50$.

QUIZ QUESTION

What is the next number that can be written as the sum of 2 squares? Hint: it's less than 100.

But 50 has even greater significance. For example, the cast of *Baywatch* go through 50 gallons of sunscreen on the set every season, and fleas can accelerate 50 times more quickly than the space shuttle. As well, as we go to press, 50 former guest stars on the 1980s hit TV show *The Love Boat* are now dead! Apparently they're dropping like flies. Sonny Bono was one of the most recent to pass on.

The highest that the most famous sci-fi TV series in history, *Star Trek*, ever rated was 50th. It was a major flop when it was 1st run between 1966 and 1969. Advertisers weren't keen on it, either, because most of its audience were children and teenagers, who were perceived to have little purchasing power. Well, the poor teenagers of yesterday are the wealthy, overweight computer programmers of today – so there you go.

The bestselling jazz musician of all time, the clarinet-playing, heavily permed Kenny G, has sold 50 million records. I am proud to say that I have done nothing to contribute to this total.

The earliest known unit of weight was the **mina**. It was created by the Babylonians and adopted by the Hittites, Phoenicians, Assyrians, Egyptians, Hebrews and Greeks. The actual weight of the mina varied from around $\frac{1}{2}$ to 1 kilogram from time to time and culture to culture. In the Ancient Hebrew system of weights, 60 shekels made 1 sacred mina, and 50 sacred minas made 1 sacred talent. Ancient cultures defined a unit of weight called a grain, which was the weight of a certain number of grain kernels, wheat, corn etc. For example, the Sumerian shekel was the weight of 180 wheat grains. The concept survives today, and the metric grain of 50 milligrams is used in weighing gems and precious stones.

The longest recorded life ever led by a termite is 50 years. How it celebrated the occasion is unknown, but I suspect it and a few mates would have found a nice restaurant and eaten it. The biggest termite in the world is in Providence, Rhode Island. I'm not sure of its age. The biggest boll weevil (in case you don't particularly like termites) is in Enterprise, Alabama.

In the 1960s TV spy spoof show *Get Smart*, there were 51 different phones over the 138 episodes. They were:

- address book phone
- axe phone
- balloon phone
- belt phone
- briefcase phone
- Bunsen burner phone
- car radiator phone
- clock phone
- comb phone
- compact phone
- daisy phone
- donkey-shoe phone
- doughnut phone
- eyeglass phone
- fingernail phone
- fireplace phone
- freezer phone
- garter phone
- golf shoe phone
- gun phone
- hair dryer phone
- handkerchief phone
- headboard phone
- headlight phone
- hose phone
- hotline to the White House
- hydrant phone
- ice-cream cone phone
- jacket sleeves phone
- lighter phone
- longhorns phone
- magazine phone
- microscope phone
- miniature phone hidden under a normal phone
- mummy phone
- 99's portrait phone
- perfume spray phone
- phone phone
- plant phone

- sandwich phone
- shepherd's staff phone
- shoe phone
- sock phone
- steering wheel phone
- test tube phone
- thermos phone
- thumbnail phone
- tie phone
- wallet phone
- watch phone
- water canteen phone

 In the year 51 BC, Cleopatra became queen of Egypt.

 Fifty-one is the 14th lucky number.

Fifty-two is an **untouchable number**: it is never the sum of the proper divisors of any other number. What the hell does that mean? Well, the proper divisors of 38 are 1, 2 and 19, and 1 + 2 + 19 = 22. So 22 is the sum of the proper divisors of 38, which means 22 is not an untouchable number. Similarly, the divisors of 60 are 1, 2, 3, 4, 5, 6, 10, 12, 15, 20 and 30, and 1 + 2 + 3 + 4 + 5 + 6 + 10 + 12 + 15 + 20 + 30 = 108. So 108 is also not an untouchable number. But no number has a list of proper divisors which add up to 52, making 52 untouchable. The only other untouchable numbers under 100 are 2, 5, 88 and 96.

The standard deck of cards used in the west has 52 cards, excluding jokers. The cards are adapted from the 56 numbered cards in the tarot deck, popular in medieval Italy. The design of the cards' faces was originally French, with the spades representing pikes (piques, in French) and the clubs the French trefoil symbol. The design of the back varies widely, from standard prints to advertising for whisky manufacturers via some truly low-grade pornographic photos. I'd strongly recommend that you inspect the back of any prospective pack well before you utter the phrase 'Happy birthday, Grandma!'

Napoleon lived to the age of 52 before passing away in 1821 on the British island of St Helena in the South Atlantic.

QUIZ QUESTION

Which mega-celebrity-cum-shopping-mall-security-guard received exactly 52 anger management sessions of therapy for punching out an autograph seeker?

A nswer me this: 'Find a number that, when you divide it by 3, leaves a remainder of 2 that, when you divide it by 5, leaves a remainder of 3 that, when divided by 7, leaves a remainder of 4.' Well, let's have a go. If you want to do most of the work yourself, cover the page and just read 1 line at a time.

- We'll start with the 3rd requirement. What sort of number, when divided by 7, leaves a remainder of 4? Clearly 4 would do, as would 11, because it equals $7 + 4$, and 18, because it equals $14 + 4$.

- The pattern is clear: a number that leaves a remainder of 4, when divided by 7, must be 7 times some number plus 4. To write a neato formula, x does the job if $x = 7k + 4$, where k is a whole number.

- And the other rules we need to satisfy are $x = 5m + 3$ and $x = 3p + 2$, where m and p are both whole numbers. (You can use k and m and p or anything you want. We just write different letters for each equation so we aren't suggesting that k, m and p have to be the same number. It's just the x that has to match up.)

- So let's solve this cow of a question by sheer hard work. The solutions to $x = 7k + 4$ are 4, 11, 18, 25, 32, 39, 46, 53, 60, 67, 74, 81, 88, 95 . . .

- Write out all the solutions to $x = 5m + 3$ and you should find only 3 numbers that match both lists. Those numbers are 18, 53 and 88.
- Now, check the 3rd rule as well by dividing each of these numbers by 3 and seeing what remainder you get. Not surprisingly, given which section we are in, the answer that satisfies all 3 rules is 53.

Fifty-three is the atomic number of iodine (I). Iodine has made liars out of many mothers when they've applied it to their scrape-kneed kids with the line: 'This won't hurt a bit.'

John Wallis, a mathematician of the 17th century, was the most influential English mathematician before Newton. He once calculated mentally the square root of a number of 53 digits, although it wasn't until a month later that he dictated the answer. This was quite typical of the kinds of things John could do, but what makes it amazing is the fact that he was 53 years old at the time. Most calculating geniuses are children.

From Algeria to Zimbabwe, via Morocco, Togo and Uganda, there are 53 countries in Africa.

And $53 = 2^2 + 7^2 = 1^2 + 4^2 + 6^2$.

What is the most annoying object ever invented? The answer, of course, is Rubik's cube, a little plastic cube consisting of several smaller cubes in six colours. Erno Rubik's 1st working prototype was made in 1974. After a lot of experimentation, Erno settled on the 3 × 3 unit cube, which resulted in 54 outer surfaces. Which makes it bloody difficult to solve.

Zacharias Dase, who was born in 1824 in Hamburg, Germany, calculated that:

79 532 853 × 93 758 479 = 7 456 879 327 810 587

Time taken: 54 seconds.

The world record for the highest stilt walk is claimed to be held by Doug Hunt, who took 54 steps on stilts 15.375 metres high (50 feet 5 inches) in Brantford, Ontario in Canada. Doug smashed Steady Eddie's previous record of 12.44 metres in height (40 feet 9½ inches). Good work, Doug. I mean, 12.44 metres – that's just a decent pair of platform shoes as far as I'm concerned.

Fifty-four is an abundant number.

Currently 54 per cent of Australians are cremated – that is, once they're dead. Also see ➤46.

QUIZ QUESTION

In 1954 Elvis Presley made his only television commercial. What was he flogging?

The most famous disco ever was Studio 54, at 254 West 54th Street in New York. Over the dance floor, tubes studded with lights rose and descended, and above them the infamous Spoon made its journeys to the nose of the Man in the Moon. Welcome to the 1970s.

QUIZ QUESTION

Who played club owner Steve Rubell in the 1998 movie *54*?

Stack a pile of tennis balls so each layer is a square. Pretty difficult, hey? But if you could, the top layer would have 1 ball, the second 4 balls, the third 9 balls, then 16 etc. The total number of balls in this pile are the **square pyramidal numbers**: 1, 5, 14, 30 . . . So 55 is the 5th square pyramidal number.

QUIZ QUESTION

What are the next 3 square pyramidal numbers after 55?

When the Apollo 12 astronauts landed on the moon in 1969, the impact caused the moon's surface to vibrate for a total of 55 minutes. The vibrations were picked up by laboratory instruments, and led to geologists theorising that the moon's surface is composed of fragile layers of rock, discrediting those who had argued for cheese.

Fifty-five is the 10th Fibonacci number and the 4th Kaprekar number. It's also the sum of all the numbers from 1 to 10. So:

$$1 + 2 + 3 + 4 + 5 + 6 + 7 + 8 + 9 + 10 = 55$$

which makes 55 the 10th triangular number. So, if you really want to intimidate someone, you could try: 'What's your favourite triangular, Kaprekar, square pyramidal, Fibonacci number? Mine would have to be 55.'

If you stack tennis balls in triangular layers, the layers will consist of 1, 3, 6 and 10 balls and so on. A triangular pyramid is a **tetrahedron**, so the tetrahedral numbers are 1, 4, 10, 20 . . . You add the number of balls in each layer together to get the numbers in the series. So:

$$1 + 3 = 4$$
$$1 + 3 + 6 = 10$$
$$1 + 3 + 6 + 10 = 20$$

etc.

Can you see that the tetrahedral numbers are just the sums of the triangular numbers? In fact, on the previous pages (➤55) you can also see that the square pyramidal numbers are the sum of the square numbers.

QUIZ QUESTION

Convince yourself that 56 is the 6th tetrahedral number.

The record for the most consecutive knockouts in a single ninja heavyweight tournament is 56. This feat was achieved by the remarkable Frank W. Dux, world champion in the late 1970s. A film based on his life, *Bloodsport* (1988), starred the Muscles from Brussels, Jean-Claude Van Damme.

Fifty-six is an abundant number.

In New York in the 1890s, German-American chef and entrepreneur Henry Heinz coined one of advertising's most famous catchphrases to describe the wide range of ketchups, sauces and relishes he'd created. The slogan '57 varieties' was a giant hit, despite the fact that Heinz actually produced 65 products at the time. Henry allegedly liked the look of 57.

The Australian dragonfly is the fastest insect that's been reliably measured on record. It has a top speed of around 57 kilometres per hour.

There are 57 letters in the longest official country name: Al Jumahiriyah al Arabiyah al Libiyah ash Shabiyah al Ishtirakiyah. No wonder we just call it 'Libya'!

The highest temperature ever recorded was on a particularly steamy day in Azizia – you guessed it, in Libya – in 1922, when the mercury topped 57°C.

The planet Mercury is almost 58 million kilometres from the sun. It hurtles through space at almost 48 kilometres per second. Venus travels at 35 kilometres per second and Earth at almost 30 kilometres per second.

The Eocene age lasted 58 million years. Splashing around at the time were lots of semi-aquatic mammals as well as reptiles, birds, fish and insects in their zillions. Homo sapiens didn't come on the scene until the Pleistocene age, 75 million years later. Late as usual but, hey, we had to do our hair.

The sum of the 1st 7 prime numbers is 58.

Baseball pitcher Orel Hershiser holds the record for the most consecutive scoreless innings thrown, in the major leagues. Orel threw down 59 no-scorers in a row, which, depending on your point of view, is either the most amazing of achievements or the most interminably boring sporting record imaginable. No offence, Orel.

 Leonhard Euler proved that:

$$635\ 318\ 657 = 59^4 + 158^4 = 133^4 + 134^4$$

That is, there is a number that is the sum of two 4th powers in two different ways.

 A day on Mercury lasts 59 Earth days.

The Greek astronomer Oenopides discovered the period of the Great Year. Originally, the Great Year was the period after which the motions of the sun and moon came to repeat themselves. Later, it came to mean the period after which the motions of the sun, moon and planets all repeated themselves – so, in the period of 1 Great Year, all should have returned to their positions for the beginning of the next Great Year. One Great Year equals 59 Earth years, so next New Year's Eve when you tell someone to have a great year, keep in mind that they mightn't actually live to see the end of it.

The average komodo dragon weighs 59 kilograms. So if you're looking at grabbing a strapless, red off-the-shoulder number for a komodo dragon mate of yours, a size 12 is probably a safe, loose-fitting bet.

The longest that any chimpanzee has jumped around, scratched its head and done those big toothy smiles is 59 years.

Among the human cannonballs of the world – who, let's be honest, we can assume to be a fairly intense group of men and women – there is one who stands supreme. The big boomer, the guru of gunpowder, the sultan of sheer silliness is Dave Smith Sr, an American projectile who has travelled a world-record 59 metres. For reference, it would take the average person about 20 seconds to walk this distance, with nowhere near the risk of injury, much less preparation time and absolutely no need to wear a silly, sequin-covered suit (unless they wanted to).

A 15-round boxing match lasts 59 minutes.

The Babylonians and Chaldeans counted in base 60 (see ➤10 if you've forgotten what that means), which is really hard to get our base-10 heads around, especially because a base-60 system is called a **sexadecimal system.** But 60 has the advantage of being easily divided by 2, 3, 4, 5 and 6 as well as 10, 12, 15, 20 and 30. The Babylonians divided a circle into 360 degrees and today we still divide hours into 60 minutes and minutes into 60 seconds.

Sixty is an abundant number, and 60 degrees is the interior angle of an equilateral triangle.

The endurance record for dart playing is allegedly held by Vincent Peter and was achieved on a date unknown and witnessed by unknowns. Not convinced? Well, we know he played for 60 hours at the Elain Town Hall in France, scored 461 880 and threw 29 292 darts. The previous record was a mere 48 hours at the Seaforth Hotel, Stornoway, Isle of Harris, by John Macaskill.

In German, 60 eggs equals a *Schock* of eggs.

Sixty-one and 59 are **twin primes**, that is, a pair of primes that differ by 2. Mathematicians have found millions of twin primes – 3 and 5, 41 and 43, 1 000 000 000 061 and 1 000 000 000 063 – and most believe that there are an infinite amount of them. No-one has yet proven it, though.

Back in the 1700s, they very rarely had the gigantic week-long parties that we call mathematics conferences. Similarly, there weren't many mathematical journals or magazines. So the main way of communicating your ideas and achievements was by writing to another mathematician. One day, Fermat banged out a letter to L'Hôpital, challenging him to find whole numbers x and y which would solve $x^2 - 61y^2 = 1$. An equation like this, where the solutions are whole numbers, is called a **Diophantine equation** (see ➤84), and Fermat ate such equations for dinner. Euler used to call this Pell's equation, which is bizarre because Pell had nothing to do with it – but that's another thing altogether. Anyway, I won't make your head explode by trying to solve it; the answer is $x = 1\ 766\ 319\ 049$ and $y = 226\ 153\ 980$. Hey, you think that's tough. Fermat also worked out that the smallest whole-number solution to $x^2 - 109y^2 = 1$ is $x = 158\ 070\ 671\ 986\ 249$ and $y = 15\ 140\ 424\ 455\ 100$.

QUIZ QUESTION

Perhaps you'd like to try to solve $x^2 - 37y^2 = 1$, with the hint that $x = 73$.

Unusually large numbers of UFO sightings occur every 61 months, according to Professor David Saunders of the Psychology Department of the University of Chicago.

American baseball fans went wild in 1998 when Mark McGwire and Sammy Sosa both hit their 62nd home runs for the season, passing the old record of 61 held by Roger Maris. American engineer Mike Keith has noted that if you add the digits of these numbers and their prime factors, the result is equal. Let's try it. Sixty-one is prime, so its only prime factor is itself. So adding the digits of 61 and its prime factors, we get $6 + 1 + 6 + 1 = 14$. But $62 = 2 \times 31$, so 2 and 31 are prime factors of 62. Adding the digits of 62 and its prime factors, we get $6 + 2 + 2 + 3 + 1 = 14$. Mike has christened such numbers Maris-McGwire-Sosa pairs. Other MMS pairs include 7 and 8, 14 and 15, 43 and 44, 50 and 51, 63 and 64, 67 and 68, 80 and 81, 84 and 85, up to 997 and 998. For more, check out the web site www.maa.org/mathland/mathtrek_9_28_98.html

Muhammad lived until he was 62 years old. He had 11 wives and 2 consorts during that time, but none of his sons survived to succeed him.

Elvis Aaron Presley, King of Rock 'n' Roll, karate-chopped and pelvically thrusted his way through a record 62 gold-selling, sequin-embossed, cheeseburger-fuelled albums.

The oldest-ever horse and the oldest-ever ostrich both made it to 62 years. As far as I know, they never met.

The least impressive mountain in the world and the world's lowest named peak is Mount Alvernia in the Bahamas, which towers a measly 63 metres above sea level. I mention this intending no disrespect to those mighty adventurers who have conquered Mount Alvernia.

Sixty-three is the 15th lucky number.

Cats and dogs are pregnant for 63 days, but rarely to each other, despite the best efforts of many contestants of *Funniest Home Videos*.

The second 6th power after 1, 64 is also a square and a cube, making it a handy number: it equals $8^2 = 4^3 = 2^6$.

There are 64 squares on a chess board. Sexy chess words include: 'chatarunga' and 'shatranj' – Indian board games from which chess probably evolved; 'Ruy Lopez' and 'anti Veresov variation' – different possible opening tactics; and, my favourite, 'zugzwang' – a situation where you are forced to move, even though any move you take will hurt you. At many times during its history, chess was banned by religious leaders and royalty, including King Louis IX, who banned the game in France in 1254. How this could have been a priority while Europe battled Mongol hordes, the crusades raged and France battled the English Plantagenet kings eludes me.

Next time you want to impress your friends, tell them that 64 is really 1 000 000. They'll stare at you in bewilderment until you explain **binary numbers** to them. We've already discussed base 10 (see ➤10) – well, binary numbers use base 2. We write 7038 the way we do because it is:

$$7 \times 10^3 + 0 \times 10^2 + 3 \times 10 + 8 \times 1$$

So to write a number in binary we break it down not into powers of 10, but into powers of 2. For example:

$$25 = 1 \times 16 + 1 \times 8 + 0 \times 4 + 0 \times 2 + 1$$

so in binary numbers we would write this as 11001. Now, 64 is a power of 2, so to write it in base 2, or binary form, we notice that:

$$64 = 1 \times 64 + 0 \times 32 + 0 \times 16 + 0 \times 8 + 0 \times 4 +$$
$$0 \times 2 + 0 \times 1$$

So 64 = 1 000 000 (in binary form).

Sixty-five is the constant of a 5 × 5 magic square. Such a square contains the numbers 1 to 25, and all the rows and columns and each diagonal will add up to 65. Hey, here are a couple:

1	7	23	20	14
18	15	4	6	22
9	21	17	13	5
12	3	10	24	16
25	19	11	2	8

1	10	22	18	14
17	13	4	6	25
9	21	20	12	3
15	2	8	24	16
23	19	11	5	7

 Jackie Chan has risked his life by appearing in more than 65 movies. Jackie does all his own stunts, and no insurance company will underwrite his movies.

 Sharks can travel up to 65 kilometres an hour (about 40 miles).

Before Madonna set a new record in *Evita* (see ➤85), the most costume changes in a Hollywood movie had been by Elizabeth Taylor, in *Cleopatra*, with 65.

And here's a fairly sexy equation:
$$65 = 1^2 + 8^2 = 4^2 + 7^2$$

Sixty-six is an abundant number. It's also a **palindromic triangular number** (see ➤15).

QUIZ QUESTION

What are the next 2 palindromic triangular numbers after 66?

The number 66 is a symbol for Allah.

Jazz Latino groove master Sergio Mendes and his band Brasil 66 had a hit in 1966 with a jazz–bossa nova version of 'Mas Que Nada'. Sergio had many bands: there was Brasil 65, Brasil 77, Brasil 88, and so on.

The King James version of the Bible has 66 books. It also has 1189 chapters and 31 173 verses. Thirty-nine books are in the Old Testament, which, if you haven't read them, are like the prequels, and 27 are in the New Testament, which is more like the actual *Star Wars* movies. (Please note: this is a very rough analogy.) If you want to impress people by quoting a couple of little-known books of the bible, throw around the names Micah from the Old Testament and Colossians from the New Testament.

Most schoolkids learn that the measurement of right-angled triangles (triangles with 90 degrees in 1 corner) gives examples of where $a^2 + b^2 = c^2$. Like this:

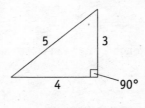

$$3^2 + 4^2 = 5^2$$

One of the most famous ideas ever put forward in mathematics came from the Frenchman Pierre de Fermat who claimed that, while there are an infinite number of examples of $a^2 + b^2 = c^2$, like the one that we see above, there are no whole numbers that will satisfy $a^3 + b^3 = c^3$ or $a^4 + b^4 = c^4$ or $a^{36} + b^{36} = c^{36}$. Indeed, there's no solution to $a^n + b^n = c^n$ for whole numbers if n is greater than 2. This problem baffled the greatest minds and only succumbed to a monstrously large proof 350 years after Pierre 1st raised the idea. For a long time, though, mathematicians had suspected that the claim was true and had pushed the limits higher and higher for the lowest possible value of n if there was to be a counterexample. Basically, they knew that $a^{5\,000\,000} + b^{5\,000\,000} = c^{5\,000\,000}$ couldn't happen, nor could it for any powers less than 5 000 000, but it took ages to bust the problem for all possible values of

n. Why am I telling you all this? Well, one of the 1st major players in trying to solve Fermat's Last Theorem (as this problem was known) was German megabrain Ernst Edward Kummer. He proved that there was no solution to $a^n + b^n = c^n$, when n was a regular prime (which, despite sounding straightforward, has a wicked definition that I don't want to bore you with here). When he'd done that, all the cases from $n = 1$ to $n = 100$ could be instantly removed, except 3 curly, irregular primes. They were $n = 37$, $n = 59$ and, you guessed it, $n = 67$.

Sixty-seven is the 16th lucky number.

Eleven cuts through a pizza can create up to 67 pieces.

On 7 May 1980 Paul Geidel walked free from the Fishkill Correctional Facility in Beacon, New York State, having served time for murder. Hey, so what, that happens every day somewhere around the world. The difference here was that Paul Geidel had entered prison on 5 September 1911, making his stay 68 years, 8 months and 2 days. That's a lot of bad mashed potato and stew.

While it is now the international home of charcoal suits with braces, gallons of gel per head and some of the worst ties and vests you'll ever lay eyes on, the world's most famous stock exchange began in 1792 underneath a tree outside number 68 Wall Street, New York, New York.

Guinea pigs are pregnant for 68 days but pack in a lot of pregnancies if given the chance, even though they live for just 3 years.

The last 68 of the 80 episodes of TV's *Dad's Army* were made in colour. The series ran from 1968 to 1977 and counted among its directors Bob Spiers, the man responsible for many classic British TV comedies – including *The Goodies*, *Not the Nine O'Clock News*, *Absolutely Fabulous* and the greatest of them all, *Fawlty Towers* (see ➤12). Isn't it amazing that someone so talented could make such a piece of crap as *Dad's Army*?

Sixty-nine is the 17th lucky number. It's also the only number whose square and cube between them use all the digits 0–9 once each: $69^2 = 4761$ and $69^3 = 328\,509$.

The French for 69 is *soixante-neuf*, and the number is loved worldwide.

Seventy is a **weird number**, but not because it's kinky or possibly mentally ill. A weird number is one that is abundant but isn't equal to the sum of any of its divisors. Weird numbers are very rare, and 70 is the only one below 100. See, 12 is abundant because $1 + 2 + 3 + 4 + 6 = 16$, which is greater than 12. But $2 + 4 + 6 = 12$, so 12 isn't weird. See if you can convince yourself that 70 is weird.

The longest game of Monopoly on record lasted 70 days. That's weird.

The Brorsen-Metcalf comet pays a visit to Earth's skies every 70 years or so. It last visited in 1989, while it previously passed closest to us in 1919. However, it was 1st discovered in the sky in 1847, 72 years previous to its 1919 visit. Which goes to show that even comets aren't always reliable.

Seventy-one is prime. About 2300 years ago the famous mathematician Euclid proved that there are an infinite number of primes, in Proposition 20 of Book IX of his giant work *The Elements*. The big guy did it this way:

- Let's say there is only a finite number of primes and that the biggest prime number is p.

- Now think about the number

$$2 \times 3 \times 4 \times \ldots \times p + 1$$

That is, the number you get when you multiply together all the numbers up to and including this biggest prime p and then add 1. Let's call it q.

- Look at this number q. It isn't divisible by 2, because when you divide it by 2 you get the remainder of 1 that comes from the + 1 at the end. But similarly, it isn't divisible by 3 or 4 or 5 or any number up to p, because each time the division will still leave a remainder of 1.

- So, 1 of 2 things must happen here. Either there is a prime number larger than p that divides into q, or q is a prime number. Either way, p can't be the biggest-ever prime. Assuming that there *is* a largest prime number leads to this inherent contradiction, so there must be an infinite number of prime numbers kicking around. To catch the largest one that's been found yet, see ➤31.

Most of us would have trouble balancing on 1 foot for a couple of minutes without tumbling over hopelessly. Get towards 15 minutes and a mixture of fatigue and sheer boredom would surely set in. So we should take a moment to applaud the effort of Amresh Kumar Jha, from Bihar in India, who has balanced on 1 foot not for 71 seconds, nor for 71 minutes, but for 71 hours.

Seventy-one per cent of Earth's surface is water. A mere 29 per cent is dry land. So, guys, there's really no excuse for not washing at least once a week, yeah? People from Kentucky particularly take note (see ➤1).

While most elephants only live to between 30 and 40 years, the longest-recorded lifetime ever had by an elephant is 71 years.

Seventy-two to the power of five (72^5), or 1 934 917 632, as you may know it, equals 19^5 + 43^5 + 46^5 + 47^5 + 67^5, and is the smallest 5th power equal to the sum of the 5 other 5th powers.

Alan, Velma, Daphne (my favourite), Shaggy and Scooby Doo were those crime-solving 'meddling kids' on 72 episodes of *Scooby Doo*, made between 1969 and 1972.

Louis XIV was the king of France for a staggering – like 'Hey, Louis. Haven't you had enough fun yet?' – 72 years. He was born in 1638 and took over the top job when his dad, named, predictably, Louis XIII, died in 1643. Louis was a big believer in the divine right of kings, which said that he had unlimited clout given to him by God and that he was answerable to God and God alone. It's pretty obvious why the divine right of kings was a popular idea with lots of kings, but not as much of a favourite with most elected officials, business people and other non-kings. Louis XIV was succeeded by his great-grandson, Louis XV, who clocked up 59 years on the throne, but had much less of an idea of what was going down.

In the 72nd episode of *The Brady Bunch*, the whole gang heads for Hawaii. Bobby finds a tiki and then Greg wipes out in a surfing competition.

It takes 72 muscles to speak a single word, so coming out with 'Um . . . um . . . yeah . . . ah' is really a waste.

The space shuttle Challenger was 73 seconds into its mission when it exploded. Or 73.137 seconds, if you wish to quote the figure from the *Report of the Presidential Commission on the Space Shuttle Challenger Accident*, but 73 is probably good enough, really.

 In the 73rd episode of *The Brady Bunch* Jan gets her hands on the tiki and a huge poisonous wolf spider gets friendly with Peter's pyjamas. Great stuff!

 The human body has about 73 kilometres of nerves.

Seventy-three is the 18th lucky number.

One theory that is constantly being proposed and refuted on the Internet and elsewhere is that, when jointly developing the compact disc, Sony and Philips agreed on it being 74 minutes long so as to accommodate Beethoven's Ninth Symphony. This symphony was allegedly the favourite tune of the wife of the Sony boss at the time, Akio Morita, and she therefore insisted that this new form of technology be large enough to accommodate Ludwig's little ditty.

The oldest person to have a sex change was 74 years old. This may seem extraordinary, but it is in fact common for people to want to have a sex change once they reach retirement age. My theory is that it's a reaction to years of wearing bad suits to work.

The longest alphabet in the world is the Cambodian, with 74 characters. Schoolteachers take note: next time some pesky 2nd-grader is complaining about learning their ABC, point out how hard it would be to learn their Khmer letters.

Muhammad Ali and George Foreman 'rumbled in the jungle' in Zaire in 1974, in a fight organised by General Mobutu – who will be remembered more for being a ruthless and oppressive dictator than for wearing funky leopard-skin hats.

In the early 1950s, Clint Eastwood signed a $75-a-week contract with Universal Studios to do walk-ons in low-budget horror flicks like *Revenge of the Creature*. He was fired when studio executives allegedly decided his Adam's apple protruded too much for him to be star material.

Seventy-five is the 19th lucky number.

According to a recent survey, 75 per cent of people who play the car radio while driving also sing along with it. The survey didn't mention what percentage of people nod their head and flick their fluffy dice to the sounds of 'doof, doof, doof'.

Seventy-six is the 9th Lucas number. And I don't mean the number of *Star Wars* prequels and sequels to come. Check ➢18 for details. It is also **automorphic**: $76^2 = 5776$, which ends in the numbers 76, making it automorphic.

One of the most famous of celestial visitors is Halley's comet. British astronomer Edmund Halley realised that the comet he and others had seen in 1682 was the same comet seen by Kepler in 1607, and that it would be back. Halley's comet was next seen in 1758 and most recently in 1986, when it was as big a letdown as the movie *The Avengers*. It reappears roughly every 76 years.

Steve McQueen's finest hour was as an American POW in *The Great Escape*, the 1963 blockbuster about the largest Allied escape attempt from a Nazi prison camp during World War II. There were 76 escapees in real life, but none of them were American POWs. But hey, where's the movie in that?

In 1976, a Los Angeles secretary officially married a 50-pound rock. The ceremony was attended by more than 20 people, at least 1 of whom, we can safely assume, was a bit of a fruit bat.

Remember factorials? If you don't, have a quick squiz at ➤24 to remind yourself. Now, it turns out that 77! + 1 is a prime number. Trust me: it has over 115 digits, so it would be an absolute bugger to work out. The numbers *n* for which *n*! + 1 is known to be prime are *n* = 1, 2, 3, 11, 27, 37, 41, 73, 77, 116, 154, 320, 340, 399, 427, 872 and the biggest currently known prime of this form, 1477! + 1.

David Lynch's *Eraserhead*, considered by many to be his greatest movie, was released in 1977. It took 7 years to make. The star, Jack Nance, played Pete Martell in Lynch's TV series *Twin Peaks*.

Efrem Zimbalist Jr, as Stu Bailey, and Roger Smith, as Jeff Spencer, trudged through the mean streets of Los Angeles in the famous private-eye TV series *77 Sunset Strip* over 205 episodes between 1958 and 1964. One guest star who stayed in the crime game was Adam West: he went on to play Batman in *Batman*.

In 1977 The Clash released a song called '1977' that ended in football-terrace chanting of '1984' (see also ➤84). I think they were trying to tell us something.

The sum of the 1st 8 prime numbers is 77.

Seventy-eight is an abundant number. It's also the 12th triangular number. Let's play with triangular numbers for a bit.

- The triangular numbers below 100 are 1, 3, 6, 10, 15, 21, 28, 36, 45, 55, 66, 78 and 91.

- Now, add any 2 consecutive triangular numbers, say, $3 + 6 = 9$, or $21 + 28 = 49$, or $66 + 78 = 144$.

- What do you notice? It seems that each time we add 2 consecutive triangular numbers, we get a square number: $9 = 3^2$, $49 = 7^2$ and $144 = 12^2$. And if you check, you'll see it happens everywhere along this list. What is happening here?

- It's most easily seen by looking at diagrams. Let's add 21 and 28 by looking at the triangular numbers: $21 + 28 = 49 = 7^2$.

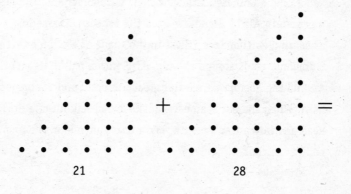

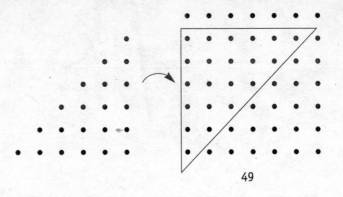

49

- So a 7 × 7 square is clearly a triangle with base 6 and a triangle with base 7 added together. This will happen any time you add 2 consecutive triangular numbers.
- If you want a sexy formula for this, write T_n for the nth triangular number and you get $T_{(n-1)} + T_n = n^2$.

The atmosphere on Earth (in its pristine state, when there's not too many fossil-fuel emissions hanging around) is 78 per cent nitrogen. There's just 21 per cent oxygen (revise that down on a bad day in Mexico City, when the air is more like gazpacho), 1 per cent argon and 0.03 per cent carbon dioxide, as well as a few traces of other gases.

Heng, a Chinese mathematician, astronomer and geographer, was born in AD 78. Heng invented the 1st seismoscope for measuring earthquakes. It was in the shape of a cylinder with 8 dragons' heads around the top, each with a ball in its mouth. Around the bottom were 8 frogs, each directly under a dragon's head. When there was an earthquake, a ball fell out of a dragon's head and into a frog's mouth, just in case anyone in the room hadn't noticed.

The inventor of the Analytical Engine, Charles Babbage, died at the age of 79. Although the Engine never progressed beyond detailed drawings, it was similar to a present-day computer. Babbage never built an operational, mechanical computer, but his design concepts have been proved correct and recently Babbage's computer was built following his own design criteria. After Babbage's death, a British committee asked to report on the feasibility of the design commented that 'its successful realisation might make an epoch in the history of computation equally memorable with that of the introduction of logarithms'.

Seventy-nine is the 20th lucky number.

Twelve cuts to a pizza can create up to 79 pieces.

Captain Kirk, Spock and the gang boldly went where no man (or woman or Vulcan, for that matter) had gone before on 79 different television occasions.

In AD 79, Mount Vesuvius erupted and destroyed the Roman city of Pompeii.

The average non-smoking, exercising flamingo with no other health problems can live to an age of 80 years.

Unlike the flamingo, the average American consumes 80 pounds of fruit and 116 pounds of beef each year.

The Roman Colosseum was completed in the year AD 80, which was good news for cats but bad news for Christians.

A variation on the magic square is the magic star of David. There are 80 magic stars of David, and each row must add up to 26.

QUIZ QUESTION

Can you solve these magic stars of David?

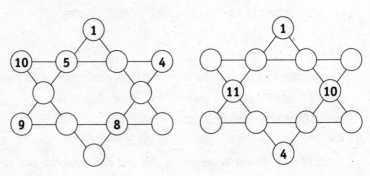

Eighty-one is the only number whose square root is equal to the sum of its digits (apart from 0 and 1). In the case of 81, $8 + 1 = 9$ and 9 is the square root of 81. It's both square and heptagonal, which somehow seems like a contradiction in terms. *But*, the fraction $\frac{1}{81} = 0.012345679\ 012345679\ 012\ldots$ and so on.

Mr Burns, the rich old man in *The Simpsons* who is Homer's boss, is usually 81 years old, except in the episode 'Who Shot Mr Burns?', when he's 104.

A ny even number greater than 68 can be written in at least 2 ways, as the sum of 2 composite odd numbers. So, for example, 82 = 33 + 49 and 27 + 55 and 57 + 25, and a few others as well. But notice that none of the odd numbers used here are prime. They are composite numbers (see ➤4). Try it yourself with any even number greater than 68. Thanks a packet to Rummler and Minnich for cracking that one.

Eighty-two is the atomic number of lead (Pb).

In 1812, 8-year-old Zerah Colburn from Vermont visited Europe to demonstrate his skills. And he had a few. He could instantly give the product of 2 numbers each of 4 digits but it took him a bit longer if both numbers exceeded 10 000. Asked for the factors of 171 395 he quickly gave 5, 7, 59 and, yup, 83. Handy trick.

Eighty-three per cent of people hit by lightning are men. The most anyone has been hit by lightning is 7 times. The lucky recipient avoided an 8th by committing suicide before it could happen.

The lovely, innocent Will Robinson, the clunky robot and the lame-as Dr Smith got into 'DANGER, DANGER' on 83 episodes of *Lost in Space*. Every one of them was better than the incredibly weak *Lost in Space* film starring Matt LeBlanc.

One of the most famous mathematicians of the ancient world was Diophantus, after whom **Diophantine equations** (equations that have whole numbers as their solutions) are named. Diophantus was so maths-mad there is even a riddle or equation in the famous *Greek Anthology* that apparently appeared on his tombstone: 'God granted him to be a boy for $\frac{1}{6}$ of this life, and adding $\frac{1}{12}$ part to this, he clothed his cheeks with down. He lit him the light of wedlock after $\frac{1}{7}$ part, and 5 years after his marriage he gave him a son. Alas, late-born wretched child! After attaining the measure of half his father's life, chill Fate took him. After consoling his grief by the study of numbers for 4 years, Diophantus ended his life.'

QUIZ QUESTION

Can you prove that the answer to this riddle is that Diophantus lived to be 84?

The book *1984* was George Orwell's 1949 indictment of Stalinism and, generally, totalitarianism (see also ➢77). David Bowie was enamoured with the idea, enough to write a song called '1984'. It appeared on *Diamond Dogs*.

Eighty-four is an abundant number.

Eighty-five can be written as the sum of 2 squares in 2 ways.

QUIZ QUESTION

Try to find the 2 ways that 85 can be written as the sum of 2 squares.

The highest recorded number of different costumes ever sashayed around a Hollywood movie was 85, by Madonna in *Evita*. Madonna also sported 39 hats, 45 pairs of shoes and 56 pairs of earrings in her groundbreaking role as Eva Peron.

Two to the power of 86 (2^{86}), when written out, does not have a zero in it. Eighty-six is almost certainly the highest number for which this happens.

Agent 86 was the code name for Maxwell Smart, the incompetent secret agent in the 1960s TV show *Get Smart*. His partner and eventual wife was Agent 99 (see also ➤51 and ➤99).

The wealthiest one-fifth of the world's population consume (or throw away, if they don't really want it) 86 per cent of its gross domestic product.

Aristarchus was a Greek astronomer who genuinely had his gear together way back around 270 BC. Using a modified sundial called a *skaphe*, he could measure both the direction and height of the sun. As a result of some pretty neat maths, he calculated that the angle between the sun, the moon and the earth was 87 degrees during a half-moon. 'So,' he decided, 'the sun must be heaps larger than the earth and miles away.' In fact it's unlikely that he said 'heaps' or 'miles' for obvious reasons. Aristarchus also thought that the earth went around the sun and not vice versa, which is pretty much accepted now, but was way controversial back then.

Ten to the power of 87 (10^{87}), that is, a 1 with 87 zeros after it, is a good estimate of the number of particles in the universe.

Eighty-seven is the 21st lucky number. Ironically, 13 is also a lucky number.

Being 13 less than 100, 87 has scared the pants off generations of cricket players who, let's be honest, must panic very easily.

With 88 or more people in a room, there's a better than even chance that 3 of them share a birthday. See ➤23.

Starsky and Hutch blowdried their hair, reached for their sunglasses and used a crowbar to get into their jeans 88 times in the 1970s.

A year on Mercury (the time it takes to orbit the sun) is equal to 88 Earth days. See ➤59.

Eighty-eight per cent of an iceberg will always be under water. And there are 88 keys on a standard piano. There was a piano on the *Titanic*. Spooky.

Try this amazing trick. Take any number, square its digits, and add them together. So 31 gives $3^2 + 1^2 = 10$, $1^2 + 0^2 = 1$; and 85 becomes $8^2 + 5^2 = 89$. Then, 89 becomes 145, which becomes 42, 20, 4, 16, 37, 58, 89. Taking any number to start and following this process, you will either descend to 1 or go to the loop that starts at 89. Try it out by starting off with a few different numbers.

Eighty-nine is the 11th Fibonacci number. Write out a long list of Fibonacci numbers and divide each number on the list by the one before it. So 1, 1, 2, 3, 5, 8, 13, 21, 34, 55, 89, 144, 233 . . . You get $1 \div 1 = 1$; $2 \div 1 = 2$; $3 \div 2 = 1.5$; $5 \div 3 = 1.666 \ldots 233 \div 144 = 1.6180555$. You'll find that this fraction gets closer and closer to 1.6180339887 – the Golden Ratio or Divine Proportion. This is one of the sexiest numbers of all – but that's another book.

The 1948 film version of Shakespeare's *Macbeth*, directed by and starring Orson Welles, originally ran for 107 minutes and featured full Scottish accents, but was cut to 89 minutes and redubbed without the accents for general release. Some of the Bard's best lines ended up on the cutting-room floor, including 'Double, double, toil and trouble', 'I have supp'd full with horrors', and the entire conversation between Macbeth and the murderers when he talks them into killing Banquo. A sequel, *Macbeth II – Macbeth Harder*, starring Bruce Willis, is not currently being considered.

Ninety is an abundant number. And the number of degrees in a right angle, which must count for something. To prove that 90 is abundant, consider this. Cows provide 90 per cent of the world's milk. And a South African bullfrog can grow to be 90 centimetres (35 inches) in length, which is longer than your arm.

Every person has nearly 400 000 radioactive atoms disintegrating into other atoms in his or her body each second. But there's no need to worry about falling apart. Each body cell contains an average of 90 trillion atoms – 225 million times that 400 000.

A recent study conducted by the Shyness Clinic in Menlo Park, California revealed that almost 90 per cent of Americans consider themselves to be shy.

A female lion does more than 90 per cent of the hunting. The male appears to be afraid to risk his life, or simply prefers to rest.

A bingo card usually has 90 numbers. Therefore there are approximately 44 million ways to make B-I-N-G-O. Ah, Grandma, now I see why it's so exciting!

Ninety-one is triangular, square pyramidal and centred hexagonal. Lucky bugger.

QUIZ QUESTION

Can you prove that 91 is triangular, square pyramidal and centred hexagonal?

Ninety-one is the number of days in a quarter-year: 13 weeks of 7 days each. And it's a pseud. I don't mean it's a pretentious, superficial wanker – no, it's a **pseudoprime**: $3^{90} - 1$ is divisible by 91, but 91 isn't prime (13 × 7 = 91).

Skippy the bush kangaroo warned of danger, carried messages and did just about everything short of performing neurosurgery while landing the helicopter on 91 episodes of *Skippy* between 1966 and 1968.

If you're wondering why you're no good at tennis, it might be because the middle of a tennis net is 91 centimetres high. Or else it's got something to do with the racquet, the balls or the alignment of the planets.

Cabbage is 91 per cent water.

Try this handy little game. Pick any number to start. Now follow this pattern:

- If the number is even, halve it.
- If the number is odd, triple it and add 1.

This is called the '3x + 1' problem, because an odd number x becomes $3x + 1$ under the 2nd rule. Let's apply the rule and see what happens. We're on this page, so let's start with 92:

- 92 is even, so we halve it and get 46.
- 46 is even, so we halve it and get 23.
- 23 is odd, so we triple it and add 1, and get 70.
- 70 is even, so we halve it and get 35.

Continue this pattern and the path is 92, 46, 23, 70, 35, 106, 53, 160, 80, 40, 20, 10, 5, 16, 8, 4, 2, 1. Now, 1 just goes 1, 4, 2, 1, 4, 2, 1 . . . so we finish at 1. It's strongly believed, but unproven, that absolutely any starting point will eventually descend to 1. Like Fermat's Last Theorem for a long time (see ➢67), millions of cases have been tested, but the general rule hasn't been proven. The problem, sometimes called the Syracuse algorithm or the Hailstone problem, is a beautifully simple one to explain, but quite baffling when you get more deeply inside it. In the 1960s it was known to tie up entire maths departments at American universities as great minds couldn't believe that such a simple process could be so difficult to understand. The main confusion comes from how differently numbers behave under the pattern. The number 29 goes 88, 44, 22, 11, 34, 17, 52, 26, 13, 40, 20, 10, 5, 16, 8, 4, 2, 1, and reaches 1 after just 18 steps with a peak of 88, while the number 31 spirals up as high as 7288 before finally reaching 1 after 106 steps. Explore the '$3x + 1$' problem yourself with different starting numbers. If you're hooked, also look at the $5x + 1$ problem and $5x + 3$, $7x + 1$ etc. You'll notice some fascinating differences.

The most forested country in the world is Surinam, which is an incredible 92 per cent woodland. Exactly what percentage of Surinam has hippies chained to it is unknown.

Ninety-two is the atomic number of uranium (U). And 13 cuts to a pizza can create up to 92 pieces. However, reheating cold pizza *with* uranium is not recommended.

The sun is made up of 93 per cent hydrogen. The rest is 5 per cent helium and 2 per cent other bits and pieces. The sun spends its time transforming hydrogen into helium, which is a full-time job.

When we don't count the elements formed by smashing others together at ridiculously high speeds, or those that exist for only 0.000 000 001 of a second before saying, 'Thanks, I'm out of here', there are only 93 naturally occurring elements.

Ninety-three is the 22nd lucky number.

Joachim Vincent Pecci was born in 1810 in Italy. On 20 February 1878, Joachim became Pope Leo XIII and kept Popeing around and doing Pope-type things until 20 July 1903, at which time he was a record 93 years old.

In the 1974 car-stealing movie *Gone in 60 Seconds* ('You can lock your car. But if he wants it . . . it's gone in 60 seconds!'), 93 cars are crashed during the 97-minute running time. Filmed in California but released only in Sweden, Finland and the Netherlands, *Gone in 60 Seconds* deserved to be in cinemas for about as long as its name suggested.

In 1742, the mathematician Christian Goldbach claimed that every even number can be written as the sum of 2 prime numbers. For example, 24 = 11 + 13 and 50 = 19 + 31. No-one has ever proven Goldbach's conjecture. It's even been shown to be correct for all even numbers up to 100 000 000 but not for every even number. So if you manage to work out why, don't forget to write it down and give me a call.

QUIZ QUESTION

Can you write 94 as the sum of 2 primes?

Americium (Am) has an atomic number of 95. All of americium's isotopes are radioactive, hence it virtually never gets invited to parties or out on dates. Isotope 243 is the most stable and is used in smoke detectors.

Because of its density, Saturn would float on water if there was an ocean large enough. However, Saturn is 95 times heavier than Earth. Go figure.

The ugly stepsisters, the pumpkin and the glass slipper have been dusted off 95 times on screen, making Cinderella the most filmed of any story. Staggeringly, to the best of my knowledge, Kevin Bacon has starred in none of these movies.

The last season of *Batman*, in 1967, began with the 96th episode, 'Enter Batgirl, Exit the Penguin'. Batgirl was played by Yvonne Craig and was actually Commissioner Gordon's daughter.

In *Measurement of the Circle*, the Greek mathematician Archimedes shows that the exact value of π lies between the values $3\frac{10}{71}$ and $3\frac{1}{7}$. He worked this out by circumscribing and inscribing a circle with regular polygons having 96 sides. Eureka!

Ninety-six is an abundant number.

The strongest any liquor can be is 190 proof. This means it's a little more than 97 per cent alcohol.

Simon Stern of Milwaukee set a record in 1984 when, at age 97, he became the oldest man ever to be divorced. Sorry to hear that it didn't work out, Simon . . . Plenty more fish in the sea.

The movie listed as 97th on the American Film Institute's list of the 100 greatest movies, which is selected by the AFI's blue-ribbon panel of more than 1500 leaders of the American movie community, is the 1938 Howard Hawks comedy *Bringing Up Baby*. In it, Katharine Hepburn, an heiress, is chasing Cary Grant, a zoologist, who is chasing a dinosaur bone stolen by her dog. They both also chase a leopard, called Baby. Trust me – it's better than it sounds.

QUIZ QUESTION

Which movie is listed as the greatest movie ever made?

There were 97 people on the Hindenburg, a German dirigible (rigid-bodied airship) that burst into flames while landing in Lakehurst, New Jersey, on 6 May 1937. Thirty-six of those people were killed.

From tip to tail, the longest recorded domestic cat was 97 centimetres long. So, if a room is 1 metre 94 centimetres or wider, strictly speaking there *is* enough room to swing a cat.

There are currently 97 countries in the world that have the death penalty.

Humans seem to share pretty much 98 per cent of their DNA with – wait for it – chimpanzees. Hi Dad, how's that chest hair going?

The most tilted planet in the solar system is Uranus, which rolls through space tilted at 98 degrees (see also ➤23). As a result, the summer of Uranus lasts up to 21 years, but for those of you calling your travel agent, note that winter goes for just as long, during which overnight temperatures can get as low as minus 219°C.

The atmosphere on Mars is about 98 per cent carbon dioxide.

The furthest that an egg has been thrown without breaking is 98 metres, in Jewett, Texas.

The world record for running a marathon is around 2 hours 6 minutes. So running a marathon in 7 hours 33 minutes is nothing exceptional, unless you happen to be Dimitrion Yordanidis, the Greek freak who did exactly that in 1976, aged 98.

Gilligan bumbled around, Ginger looked exceptionally well presented for a castaway and the Professor could build radios, weapons, houses and do just about everything else other than FIX THE BLOODY BOAT for 98 episodes of *Gilligan's Island*. This does not include the TV spectacular *Baywatch Meets Gilligan's Island*, which must rank as one of the worst pieces of television I've been unfortunate enough to witness, and believe me, I've seen some turkeys in my time.

Agent 99 was Maxwell Smart's offsider in *Get Smart*. She was played by Barbara Feldon, and was about as cool and groovy as you can get. *Get Smart* was originally written by Mel Brooks with Buck Henry. See also ➤51 and ➤86.

Ninety-nine is the 5th Kaprekar number:
$99^2 = 9801$ and $98 + 01 = 99$.

Muslim rosaries may contain 99 pearls, which represent the 99 names of Allah and the 99 names of Muhammad.

According to *Time* magazine, a staggering 99 per cent of Finnish 18–24-year-olds own a mobile phone. Setting one off in a room of twentysomethings must lead to an orgy of chest padding, pocket fossicking and frantic calls in Finnish for 'Is that me?'

If Barbie were life-size, her measurements would be, in centimetres, 99-58-83. In inches that's 39-23-33, which is absolutely perverse. She'd stand about 218 centimetres (or 7 feet 2 inches) tall, but it's highly doubtful that such a body could be supported by human feet as proportionally small as hers.

Ninety-nine per cent of Earth's water is undrinkable. But see ➤71.

If you leave a photocopier on 99 copies, it will genuinely piss off at least 1 person.

Amaze your friends with this fact:
$$99 = 2^3 + 3^3 + 4^3$$

The number 1 with 100 zeros after it is called a 'googol', a name coined by American mathematician Edward Kasner. This leads to the number '1 googolplex', which is 1 followed by a googol zeros, which is unnecessarily large and quite silly in some ways.

The 100th episode of the 1980s TV series *MacGyver* was called 'Deep Cover'. In it, Pete and Mac try to help when a Phoenix Foundation engineer is seduced and betrayed by a spy seeking her prototype for a new stealth technology for submarines. Sounds pretty . . . 1980s to me. MacGyver went on wielding his penknife for another 39 episodes.

There is a relationship between triangular numbers and cubes. Take a triangular number and square it; it is equal to the sum of some cubes. Put more formally, $(T_n)^2 = 1^3 + 2^3 + \ldots + n^3$. Now, as I mentioned earlier, 10 is the 4th triangular number, because $10 = 1 + 2 + 3 + 4$. So in the formula above, $T_4 = 10$, and we can write $10^2 = 1^3 + 2^3 + \ldots + 4^3$ or $100 = 1^3 + 2^3 + 3^3 + 4^3$.

QUIZ QUESTION

Using only $+$, $-$ or $\times$, join the digits 1–9 to make a total of 100. One way is $1 + 2 + 3 + 4 + 5 + 6 + 7 + (8 \times 9) = 100$. But there are others. Find another one.

On 11 October 1978, ex-Sex Pistol Sid Vicious called the police from room 100 of New York's famous Chelsea Hotel to say that someone had stabbed his girlfriend, Nancy Spungen. Sid was charged with the murder, but died while he was out on bail.

There are 100 tiles in the standard Scrabble set. I'd like to take this opportunity to say that Scrabble is absolutely cool, but that 3-D Scrabble, Up-words etc. suck big time. There are also 100 squares on a snakes and ladders board.

In the animal world, 100 is the number of teeth on the Hirudo leech's jaw. Oh, and it's got 3 jaws. Male sealions may mate with more than 100 females.

According to *Time* magazine, 30 August 1999, there are 100 times as many fungal spores present in ballet dancers' shoes as there are in joggers' shoes. I mention this not in any way as a payout on ballet dancers, but merely out of a desire to write the words 'fungal spores'. Ooooooohhh goody – that was great. I can go now. Bye.

QUIZ ANSWERS

1: 'Cartman Gets an Anal Probe'.

3: 354 and 138 624 147; there's the cute one with the eyes, the big silly one without eyes, and the strangely alluring Chewbacca Mr Hat.

5: Twelve of the buggers:

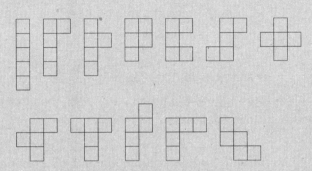

6: 28 and 496. Hey, so is 8128. Trust me.

7: $2^5 - 1 = 31$; $2^7 - 1 = 127$

9:

4	9	2
3	5	7
8	1	6

12: 18, 20, 24, 30 and 36.

15: 21, 28 and 36.

19: 209, 418 and 817.

22: 35, 51 and 70; 11 and 26.

25: 31, 33, 37, 43 and 49.

27: $27 = 3^2 + 3^2 + 3^2 = 5^2 + 1^2 + 1^2$

29: 10. Or you could just order in some more pizza.

34:

12	13	1	8
6	3	15	10
7	2	14	11
9	16	4	5

1	2	15	16
13	14	3	4
12	7	10	5
8	11	6	9

10	16	1	7
6	9	8	11
3	4	13	14
15	5	12	2

3	15	2	14
13	4	9	8
12	5	16	1
6	10	7	11

36: 1, 2, 4, 6, 12 and 24.

37: 37, 61 and 91.

38:

42: 10, (9, 1), (8, 1, 1), (8, 2), (7, 1, 1, 1),
(7, 2, 1), (7, 3) . . . right down to
(1, 1, 1, 1, 1, 1, 1, 1, 1, 1).

43: 2, 3, 5, 7, 11, 13, 17, 19, 23, 29, 31, 37, 41, 43, 47, 53, 59, 61,
67, 71, 73, 79, 83, 89 and 97.

45: 55, 99 and 297.

47: 26, 28, 36, 38, 47, 48, 53, 57, 62, 69, 72, 77, 82, 87, 94, 96
and 99.

48: 140 and 195; 1575 and 1648.

49: 444 444 888 889.

50: $65 = 1^2 + 8^2 = 4^2 + 7^2$

52: Gary Coleman.

54: Mike Myers; doughnuts.

55: 91, 140 and 204.

56: $1 + 3 + 6 + 10 + 15 + 21 = 56$

61: $y = 12$

66: 171 and 595.

80:

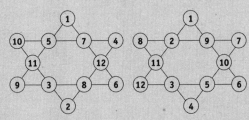

84: Let the length of Diophantus' life be called x. The riddle tells us that:

$$\tfrac{1}{6}x + \tfrac{1}{12}x + \tfrac{1}{7}x + 5 + \tfrac{1}{2}x + 4 = x$$

85: $85 = 9^2 + 2^2 = 7^2 + 6^2$

91: See ➤15 for how to work out triangular numbers, ➤55 for square pyramidal numbers and ➤37 for centred hexagonal numbers. Then go for it.

94: $94 = 41 + 53$, $5 + 89$, and many other ways.

97: Surprise, surprise – *Citizen Kane*.

100: One answer is $123 - 45 - 67 + 89 = 100$. Can you get any more?

REFERENCES

Ash, Russell. *Factastic Book of 1001 Lists*. Dorling Kindersley, Sydney, 1999.

Ash, Russell. *Top 10 of Everything 1999*. Dorling Kindersley, Sydney, 1999.

Becker, Udo (ed.). *The Continuum Encyclopedia of Symbols*. Continuum, New York, 1994.

Bowker, John (ed.). *The Oxford Dictionary of World Religions*. Oxford University Press, Oxford, 1997.

Brewer-Giorgio, Gail. *Elvis Is Alive!* Anonymous Press, New York, 1997.

Crystal, David (ed.). *The Cambridge Factfinder*. 3rd edn. Cambridge University Press, Cambridge, 1998.

Duncan, David Ewing. *The Calendar*. Fourth Estate, London, 1999.

The Guinness Book of Records 1999. Guinness Publishing, New York, 1999.

Hoffman, Paul. *The Man Who Loved Only Numbers*, Fourth Estate, London, 1998.

Kureishi, Hanif & Savage, Jon (eds). *The Faber Book of Pop*. Faber & Faber, London, 1995.

Mould, Richard F. *Mould's Medical Anecdotes*. Adam Hilger Ltd, Techno House, 1984.

Panati, Charles. *Extraordinary Origins of Everyday Things*, Harper & Row, London, 1987.

Park, James (ed.). *Icons*. Bloomsbury, London, 1991.

The Q Book of Punk Legends. Q and Guinness Publishing, Enfield, 1996.

Schimmel, Annemarie. *The Mystery of Numbers*. Oxford University Press, New York, 1993.

Stern, Jane & Stern, Michael. *Encyclopedia of Pop Culture*.
HarperCollins, New York, 1992.

Wells, David. *The Penguin Dictionary of Curious and Interesting Numbers*. Rev. edn, Penguin, Harmondsworth, 1997.

Williams, Barry & Kreski, Chris. *Growing up Brady*. HarperCollins, New York, 1992.

WEB SITES

Absolute Trivia: www.absolutetrivia.com

Ask Jeeves: www.askjeeves.com – a really good, easy-to-use search engine where humans rather than computers have indexed Internet sites, so listings are much more likely to be relevant.

Brian's Brain: http://members.tripod.com/ brainofbrian – good for useless facts, although their veracity cannot be guaranteed.

CDNOW: www.cdnow.com – fantastic for information on any kind of music, including discographies, biographies and song titles.

Encyclopaedia Britannica: www.eb.com – the entire encyclopedia online plus a useful Internet search engine that lists sites that have been read and approved by Britannica's editors.

***Get Smart* Home Page:** www.wouldyoubelieve.com – the page that lists the 51 phones so far discovered on *Get Smart*.

A Great Link: www.agreatlink.com – links to bands, celebrities, movies, TV shows etc.

Internet Movie Database: http://us.imdb.com – information on any actor or movie you can think of.

MacTutor History of Mathematics Archive: www-groups.dcs.st-and.ac.uk/~history/ – an amazing collection of biographies of great mathematicians.

Music Trivia: www.primate.wisc.edu/people/hamel/ musictriv.html – a good list of music trivia links.

Rekord-Klub Saxonia: www.recordholders.org/en/ index.html – great links to various record-breakers' sites.

Rockmine: www.rockmine.music.co.uk – a huge and fabulous site full of info on rock music, if a little UK-centric.

Topics in Mathematics: http://archives.math.utk.edu – a huge but easy-to-use index for maths sites on the Web.

INDEX